THE
BIGGEST
IDEAS IN THE
UNIVERSE

ALSO BY SEAN CARROLL

Something Deeply Hidden
The Big Picture
The Particle at the End of the Universe
From Eternity to Here

THE
BIGGEST
IDEAS IN THE
UNIVERSE

SPACE,
TIME, AND
MOTION

SEAN CARROLL

DUTTON

DUTTON

An imprint of Penguin Random House LLC
penguinrandomhouse.com

Copyright © 2022 by Sean Carroll
Penguin Random House supports copyright. Copyright fuels creativity, encourages diverse
voices, promotes free speech, and creates a vibrant culture. Thank you for
buying an authorized edition of this book and for complying with copyright
laws by not reproducing, scanning, or distributing any part of it in any form
without permission. You are supporting writers and allowing
Penguin Random House to continue to publish books for every reader.

DUTTON and the D colophon are registered trademarks of
Penguin Random House LLC.

Image on page 196 courtesy of Wikimedia Commons.
All other illustrations by Jason Torchinsky and Sean Carroll.

LIBRARY OF CONGRESS CATALOGING-IN-PUBLICATION DATA
has been applied for.

ISBN 9780593186589 (hardcover)
ISBN 9780593186596 (ebook)

Printed in the United States of America

1st Printing

BOOK DESIGN BY TIFFANY ESTREICHER

While the author has made every effort to provide accurate telephone
numbers, internet addresses, and other contact information at the time of
publication, neither the publisher nor the author assumes any responsibility
for errors or for changes that occur after publication. Further, the publisher
does not have any control over and does not assume any responsibility
for author or third-party websites or their content.

To Jennifer

CONTENTS

INTRODUCTION

My dream is to live in a world where most people have informed views and passionate opinions about modern physics. Where you knock off after a hard day at work, head down to the pub with friends, and argue over your favorite dark-matter candidate, or competing interpretations of quantum mechanics. A world where, as kids are running around at a birthday party, one parent says, "I don't see why anyone thinks there should be new particles near the electroweak scale," and another immediately replies, "Then how in the world are you going to address the hierarchy problem?" People have opinions, after all, about supply-side economics or critical race theory. Why not inflationary cosmology and superstring theory?

That's not quite the world in which we live. Even more than most other academic disciplines, physics is a field by and for specialists. Practitioners talk to one another in a highly specialized jargon, one that is dominated by mathematical concepts most people have never heard of, much less mastered. There are sensible reasons why this is the case, but it doesn't have to be this way. The situation is due in large part to the ways in which physicists tend to share their knowledge with the rest of the world.

If you are a non-expert interested in learning about modern physics, you have basically two options. One is to remain at a popular level of explanation, where you can learn about some of the relevant concepts without digging into the technical or mathematical details. You can read books, go to lectures, watch videos, listen to podcasts. The good news is that we do have a vibrant ecosystem of such resources, and it's possible to learn quite a bit, albeit in a somewhat haphazard way. But at the end, you know you're not getting the *real* stuff. What you get are images and metaphors, rough translations of the underlying mathematical essence into ordinary language. You can go an impressive distance down this route, but something vital will always be missing.

The other route is to become a physics student. That could be literally at a university, or by assembling the right textbooks and online resources. Along the way you will need to become proficient at quite a bit of mathematics: calculus and differential equations most importantly, but also aspects of vector analysis, complex numbers, linear algebra, and more. The journey will be rewarding, but frustratingly slow. It typically takes at least a year of introductory courses before a student ever hears about relativity or quantum mechanics. And most physics students can get an undergraduate degree—or even go all the way to obtain a PhD—without learning about particle physics, black holes, or cosmology. Those goodies are reserved only for specialists in particular subfields.

The gap between learning physics as an interested amateur, relying on metaphors and murky translations, and becoming a credentialed expert, comfortable with pushing around equations of intimidating complexity, is wide but not unpassable. Just because I don't want to be a professional race-car driver doesn't mean I shouldn't be allowed to drive at all. Surely there is a way to engage with some of the authentic essence of modern physics—even if that means looking at a few equations—without slogging through years of a standard curriculum.

You've come to the right place!

The Biggest Ideas in the Universe is dedicated to the idea that it is possible to learn about modern physics for real, equations and all, even if you are more amateur than professional and have every intention of staying that way. It is meant for people who have no more mathematical experience than high school algebra, but are willing to look at an equation and think about what it means. If you're willing to do that bit of thinking, a new world opens up.

Here is the thing about equations: They are not that scary. They are just a way to compactly summarize a relationship between different quantities. It's one thing to be told that, according to Einstein's theory of general relativity, "mass and energy cause spacetime to curve." It is quite another to be given Einstein's equation:

$$R_{\mu\nu} - \frac{1}{2} R g_{\mu\nu} = 8\pi G T_{\mu\nu}.$$

The English-language sentence gives you a kind of feeling for what general relativity is about, but the equation tells you what is really going on, in precise and unambiguous terms. You can read all of the words you like, but until you understand this equation, you won't really understand Einstein's theory.

The problem is that this equation is utterly opaque if you don't know what the symbols mean. It's gibberish. To wrap your head around it, you need to understand the individual roles of all the numbers and letters, including the subscripts μ and ν, which are from the Greek alphabet, for goodness' sake. There are good reasons why typical physics students take years to make it that far.

But you will get that far by reading this book. By the time you reach Chapter 8, you will understand what all the symbols in Einstein's equation mean, how they fit together, and what they are telling us about spacetime and gravity. The equation might involve Greek letters, but coming to understand it is enormously easier than, say, learning to speak and write actual Greek.

Most popular books assume that you don't want to make the effort to follow the equations. Textbooks, on the other hand, assume that you don't want to just understand the equations, you want to *solve* them. And solving these equations is enormously more work and requires enormously more practice and learning than "merely" understanding them does.

Let me elaborate on this solving/understanding distinction, because it will be the key to the remarkably fast progress we'll be able to make. Einstein's equation doesn't just relate some specific collection of mass and energy to the curvature of some specific spacetime. It is a completely general relationship, of the form "you give me some distribution of mass and energy, and I will tell you how spacetime curves in response to that." Carrying out this promise is what we mean by "solving the equation."

Sometimes solving an equation is easy: If the equation is $x = y^2$, and we're told that $y = 2$, the solution is $x = 4$. Not so hard. But real-world physics equations are more complicated than that, involving ideas from calculus (the mathematics of continuous change) and other advanced concepts. Solving such equations can become a full-time occupation for working physicists. Therefore, sensibly enough, their education consists in large part in learning to solve equations. Any physics student will tell you that the difficult part of their years in school isn't going to the lectures, it's doing the problem sets that professors keep handing out as if the students have nothing else to do that weekend.

Here in *The Biggest Ideas in the Universe*, we're not going to teach you how to solve the equations. But you will learn to *understand* the equations, even ones that are considered relatively advanced by physics-textbook standards. That turns out to be enormously easier. These books are dedicated to the belief that the ideas of modern physics—the real ones, not watered-down metaphorical versions—can be accessible to anyone willing to do just a little bit of thinking about the equations and what they mean.

———

Okay, but what are these ideas that we're talking about? There are many of them, as you might imagine. Enough that we've divided the material into a three-part series: *Space, Time, and Motion*; *Quanta and Fields*; and *Complexity and Emergence*. The trilogy format has proven successful for The Lord of the Rings and other popular franchises.

The book you hold in your hands, *Space, Time, and Motion*, focuses on the framework of classical physics pioneered by Isaac Newton, which held sway until the quantum revolution of the twentieth century. But do not fear, we will not be spending too much time on pulleys and inclined planes, as important as those are. The ambit of classical physics includes deep questions about the nature of space, time, and change, and we won't be afraid to sprinkle some philosophical considerations in among our equations. It also includes the theory of relativity, all the way up to Einstein's ideas about curved spacetime, and consequences such as black holes. So this book starts with ideas that are centuries old but will take us right up to modern research-level concepts.

In *Quanta and Fields* we will discuss sexy quantum ideas like entanglement and Schrödinger's cat, but mostly we will take the opportunity to learn about quantum field theory and particle physics, the best modern take on the fundamental laws of nature. The final installment, *Complexity and Emergence*, is where we admit that the world isn't made of just two or three particles. Interesting things happen when systems consist of a large number of moving parts.

That's a lot of concepts. And yet, almost all of them are within the realm of physics and related areas. This is not to disparage the equally big and important ideas from other areas of science (or the arts and humanities, for that matter), but one has to draw the line somewhere.

Another place we've drawn the line is between "ideas we have good reason to believe are true" and "promising speculations." While physics textbooks tend to stick to ideas that have established their usefulness, popular-level treatments will cheerfully dive into concepts that are still entirely hypothetical. That's a perfectly sensible thing to do; researchers spend most of their time at the frontier, thinking about possibilities that haven't become part of settled lore. Our goal here is to stick to ideas that we have excellent reason to think will still be part of the working physicist's tool kit a hundred years from now.

———

It is a pleasure to acknowledge the enormous help I have received along the way. Scott Aaronson, Justin Clarke-Doane, and Matt Strassler provided invaluable feedback and saved me from more than a few infelicities of expression. Jason Torchinsky made the beautiful illustrations. My editor, Stephen Morrow, was supportive and insightful as always, and my agent, Katinka Matson, helped shape the form of a complicated project. Alice Dalrymple, Tiffany Estreicher, Dora Mak, Nakeesha Warner, and Melanie Muto were crucial in the production process. The book grew out of a series of videos I made during the COVID-19 pandemic, inspired by online playwriting classes given by my friend Lauren Gunderson. And of course I cannot properly express my gratitude to Jennifer Ouellette for writing advice, moral support, and more.

To see the videos, as well as find other supplementary materials, visit:

https://preposterousuniverse.com/biggestideas/

ONE

CONSERVATION

Look around. If you're like most people, you have a body. It's located somewhere. Chances are that you are surrounded by a variety of other objects, located other places. Tables, chairs, a floor, ceiling, walls, maybe trees or a body of water if you're outside. All of these objects exist, with certain locations and properties, and those locations and properties can change with time. You can scoot your chair nearer to a wall, or farther away. You drink a glass of water, absorbing its substance into your body. If instead you put the glass on a table and leave it there, the water will eventually evaporate into the air.

That's how we think about the world from an immediate, human-scale perspective. There is stuff, which is located in space. (By "space" we don't mean "outer space," just the three-dimensional realm through which things move.) This stuff might change, or it might remain constant over time. Physics is the study of all that stuff, and its behavior, at the most basic level we can think of. What is all that stuff, really? How do different objects relate to one another? How do they change with time? What is "time," and for that matter what is "space," when you get right down to it?

One of the most enjoyable features of physics is how quickly we go from mundane observations—look at that stuff, behaving in that way!—to profound questions about the nature of reality. The key is that things don't just happen—all of the happenings fit into certain *patterns*. It's those patterns that we call the **laws of physics**, and our job is to uncover them.

The simplest pattern of all is the fact that certain things remain constant even as time passes. Contemplating that basic feature of reality is a great jumping-off point for our investigations, which will get pretty wild soon enough.

PREDICTABILITY

We take for granted that the world around us is at least a little bit predictable. If there is a table in a room, and we turn to face away from it for just a second, we expect the table to still be there when we turn back. If we place an apple on the table, we expect the table to support it, rather than the apple falling right through to the floor. As much as we might lament how difficult it is to predict the weather or future election outcomes, we should be impressed by how much reliable predictability there is.

Physics is made possible by this predictability. It may not be absolute, but we can somewhat anticipate what's going to come next in the world if we know what's going on right now. The most basic kind of predictability is **conservation**, the fact that some things don't change at all.

Conservation is just how physicists refer to "staying constant over time." You may have heard that **energy** is conserved, for example. Energy isn't a kind of substance, like water or dirt. It's a *property* that things have, depending on what they are and what kind of situation they're in. There is no "energy fluid" that flows from place to place. There are simply objects that have positions and velocities and other properties, and we can associate a certain amount of energy with them because of those facts.

An object can have energy because it is moving, because it's located at a high elevation, because it's hot, because it's massive, because it's electrically charged, or for other reasons. Under the right circumstances, those forms of energy can be converted back and forth between each other. The energy that a wineglass has just from being located on a table can, if the glass is knocked off the edge, rapidly be converted into energy of motion as it falls, and then into heat and noise and other forms of dissipated energy as it breaks on the floor. Conservation of energy is simply the idea that the total energy, given by adding up all the individual forms, remains constant throughout the whole process.

(Wait—is this circular reasoning? Are we merely inventing a bunch of quantities that add up to a constant number by definition, calling that "energy," and congratulating ourselves for discovering a law of physics? No. There is an independent way to define energy and then show that it's conserved, based on the fact that the laws of physics don't themselves change over time. But you're asking the right kind of question.)

As simple an idea as we can imagine—there is a quantity that doesn't change, it stays the same as time passes. But conservation of energy and other quantities isn't just a gentle, unintimidating place from which to launch a survey of all of physics. It's logically the right place, since an understanding of conservation was the first step in the transition from pre-modern to modern science.

FROM NATURES TO PATTERNS

Put yourself in the mindset of humans trying to understand the world before physics in its modern form came along. The Greek philosopher Aristotle is usually chosen as an exemplar, though other ancient thinkers would have thought similarly. To greatly simplify a complex and subtle set of ideas, Aristotle separated the way things move into "natural" and "unnatural" (or "violent") motions. He thought of the world as fundamentally teleological—oriented toward a future goal. Objects have natural places to be or conditions to be in, and they tend to move to those places. A rock will fall to the ground and sit there; fire will rise to the heavens.

Here on Earth, in Aristotle's view, if everything were in its natural state, things would be motionless. It requires some external influence to get things moving, and even then the motion will only be temporary. You can pick up a rock and throw it; that's an unnatural or violent motion. But eventually the rock will come back down, maybe bounce around a bit, and return to its natural state at rest on the ground.

He's not wrong, at least in a wide variety of circumstances. If you're sitting with a coffee cup on the table in front of you, by itself the cup will just sit there. You can make it move by pushing it, but when you stop pushing, it will come to rest again. We can extrapolate this, Aristotle imagines, to a basic feature of the universe. Objects are naturally at rest, and motion only occurs when something pushes them away from this natural state.

This picture fits less well with other cases that were known even in Aristotle's day. Ancient Greeks were well acquainted with arrows flying through the air. The initial force may be applied by the bowstring, but it's clear that the arrow keeps going long after it has left the bow. Why doesn't the arrow just fall to the ground? What keeps it from expeditiously returning to its natural state?

This was a question that great minds puzzled over for hundreds of years. It took a while, but the answer ultimately led to a wholesale overthrow of Aristotle's teleological view of the universe. It was replaced with a picture in which objects don't evolve toward ultimate goals; instead, they obey laws that predict what will happen the very next moment based on what's happening now.

CONSERVATION OF MOMENTUM

An important step was taken by John Philoponus, an Alexandrian thinker in the sixth century. He suggested that the bowstring imparted a certain quantity, later dubbed "impetus," to the arrow, which kept it moving for a while before eventually dissipating away. A simple suggestion, perhaps, but an important move from thinking in terms of forward-looking purposes and replacing them with properties that exist in the moment.

Philoponus's idea was developed further by Ibn Sīnā (Avicenna), an eleventh-century Persian polymath. It was Ibn Sīnā who took the crucial step of arguing that impetus is not just temporary. Every object has a certain amount of impetus (equal to zero for a stationary object, some larger number if the object is moving), and that amount remains constant unless a force somehow pushes on it.

In this new picture, the reason why rocks and coffee cups come to rest is not because that's their natural state; it's because forces—friction, air resistance—gradually degrade the impetus from the body. In the vacuum of empty space, Ibn Sīnā suggested, there would be no air resistance, and a moving body would keep moving at a constant velocity in perpetuity. This was a wildly speculative thought experiment one thousand years ago, but today we regularly build spacecraft that move between the planets at basically a constant velocity (apart from the gentle tug of gravity). In the fourteenth century, French philosopher Jean Buridan proposed a mathematical formula for the impetus, equating it to the weight of an object times its speed.

What we have here is the birth of a law of physics: **conservation of momentum**. The rough idea of some "quantity of motion" being conserved came along before anyone could pinpoint precisely what that quantity was. This is a standard story of progress in theoretical physics: We put forward a new concept, work to characterize it in quantitative terms, then take that quantitative expression—an equation—and ask how it comports with other phenomena we observe in the world. Today we know that momentum is mass times velocity (at least until relativity comes along and complicates things a bit).

One problem with Buridan's definition of impetus as "weight times speed" is that "weight" isn't an intrinsic property of an object, because it depends on the amount of gravity pulling on it—your weight would be lower on the moon than it is on Earth, and you would be weightless if you were in a spaceship coasting between the planets. **Mass**, on the other hand, is an intrinsic property; roughly speaking, mass is the resistance that an object has to being accelerated. It takes a lot of force to accelerate a high-mass object to a certain speed, and only a little force to accelerate a low-mass object to that speed.

Similarly, **speed** and **velocity** are subtly different. Speed is a number, a certain number of meters per second. Whereas velocity is a **vector**—a quantity with both a magnitude and a direction. In fact, the magnitude of the velocity vector is precisely what we call "speed," but the velocity also points in some specified direction. So you have the same speed if you're driving north at 90 km/hour as you do if you're driving south at 90 km/hour, but your velocity is different.

We denote vectors by drawing little arrows over an appropriate symbol, so the velocity of an object is typically written $\vec{v}$. We very often care about the size, or magnitude, of a vector, which is written with the same symbol but without the arrow: The magnitude of a vector $\vec{v}$ is simply v.

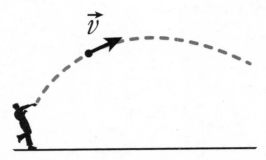

The arrow notation makes sense because we often represent a vectorial quantity by literally drawing an arrow that points in the direction of the vector, and whose length is proportional to the magnitude of the vector. Alternatively, we can represent a vector in terms of its **components**—the contributions it gets from different directions. If you are traveling exactly northward, the component of your velocity in the east/west direction is zero.

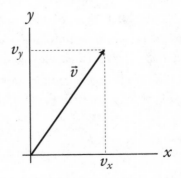

It is easy to add vectors together. Just imagine placing the beginning of the second vector at the end of the first, so we define a third resulting vector by traveling down the first and then the second. If the two vectors we're adding together point (almost) along the same direction, the total vector will be (almost) as long as the sum of their magnitudes, but if they point in (almost) opposite directions, the resulting vector can be much shorter.

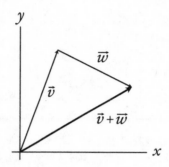

Buridan and his predecessors didn't think in terms of vectors, which were gradually developed by a number of thinkers in the nineteenth century, including German mathematician August Ferdinand Möbius (of "Möbius strip" fame); Irish mathematician William Rowan Hamilton; German polymath Hermann Grassmann; and English mathematician Oliver Heaviside. So it's no surprise that it took a while to get the right definition of momentum.

These days the momentum vector is usually denoted $\vec{p}$. (The letter m is reserved for mass, so we take the symbol from the Latin word for momentum, *petere*.) With all that in mind, the expression for momentum is the simplest thing in the world:

$$\vec{p} = m\vec{v}. \tag{1.1}$$

Our first official equation. The momentum vector points in the same direction as the velocity vector, and their magnitudes are proportional. **Proportionality** will be a crucial concept for us: It means that a multiplicative change in one quantity implies a multiplicative change in the other. If you double the velocity, you double the momentum. The factor relating the two is called the "constant of proportionality," although in some equations it might not actually be constant. In this case it is: It's just the mass of the object.

The power of even a basic equation like this should be evident. We're not saying that the momentum of some particular object just happens to be equal to its mass times its velocity; we're saying that there is a *universal* relationship between momentum, mass, and velocity, which always takes precisely this form for every object. When relativity comes along, some of the explicit forms of the equations we'll see here are going to have to be tweaked, but the basic principles are largely the same.

An equation like this has no "causality" built into it; it's a rigid relationship between the quantities involved, and it reads equally well left-to-right or right-to-left. We can manipulate the equation in any way that does the same operation to both sides, such as dividing by m to get $\vec{p}/m = \vec{v}$. We can therefore say, "If I know the velocity of an object, I multiply by its mass to get the momentum," or equally well, "If I know the momentum of an object, I divide by its mass to get the velocity."

BUMPS AND PUSHES

The power of the law of conservation of momentum goes well beyond the idea that a single object with no forces acting on it will keep moving at a constant velocity. When a force does act on an object, for example, by bumping into another object, the total momentum of the entire system remains conserved.

Imagine we have two objects that are moving without any forces acting on them, such as two billiard balls on a frictionless table. (A puck on an air-hockey table is slightly more realistic, but any invocation of "frictionless" is going to be an idealization, much as physicists love to play that card.) They are initially moving on straight lines, but then they collide and move apart on new straight trajectories.

Let's label the initial momenta (plural of momentum) of the first ball as $\vec{p}_1$(initial) and that of the second ball as $\vec{p}_2$ (initial). Following an obvious pattern, we call the final momentum of the first ball $\vec{p}_1$(final) and that of the second ball $\vec{p}_2$ (final). Then the statement of conservation of momentum in this case is

$$\vec{p}_1\left(\text{initial}\right) + \vec{p}_2\left(\text{initial}\right) = \vec{p}_1\left(\text{final}\right) + \vec{p}_2\left(\text{final}\right). \qquad (1.2)$$

The individual momenta clearly change as the balls ricochet off of each other. But the total momentum of the system as a whole remains conserved.

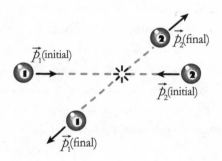

Momentum is always conserved, but we can forgive our predecessors for not noticing it right away. Think back to our Aristotelian coffee cup, which is at first motionless, then moves a bit when you push it, then comes to rest again when you stop. It looks like momentum is not conserved, because the cup goes from zero velocity to nonzero velocity. But secretly, the cup is reacting back on you, and your own body is pushing against the chair you are sitting in, which pushes against the Earth itself. The change in momentum of the coffee cup is exactly compensated by a change in momentum in the other direction in the you/chair/Earth system. You never notice, because the mass of all that stuff is so enormously large that the change in velocity is unmeasurably small, though not quite zero. When you stop pushing and the cup comes to a stop, it's because the table it's sitting on has pushed back on it, and this time the table/Earth system gains back a little momentum, returning to where it was before you started pushing anything.

In the film *Gravity*, there is a scene where Sandra Bullock and George Clooney are in peril, floating in their spacesuits outside a space station. (Spoiler alert for this paragraph and the next.) They are hanging on to the station by a single tether. In an otherwise delightful movie, the physics in this scene gets a bit wonky. What we see is that Bullock has to hold on for dear life as Clooney is being pulled away from her. In reality, since they are all in pretty much the same orbit around the Earth, there would be no forces pulling them away from the station; a gentle tug on the tether would bring them home safely. But in the movie, she has to let go of him to allow herself to return.

I don't object when filmmakers bend the laws of physics a bit in the service of a good story. But in this case, the bending was unnecessary. All you had to do to concoct an equally dramatic scene was to eliminate the tether entirely. Imagine that Bullock and Clooney had been holding on to each other but floating slowly away from the station with nothing connecting them to it at all. Conservation of momentum would tell them exactly what to do. If they stayed together they would have drifted inexorably away, and eventually both would die. But there's another option: One of them could have pushed the other one away. Their total momentum would be unchanged, but one of them could have drifted back toward the station while the other moved away even faster. Clooney could have sacrificed himself by pushing Bullock back home. Or either Clooney or Bullock could have saved themselves by pushing the other away to an inevitable doom. But that would have been a different movie.

CLASSICAL MECHANICS

Conservation of momentum remains an important physical principle to this day, but the long road to understanding it played an even more important role by introducing a new way of thinking about physics. Gone was the teleological Aristotelian world of intrinsic natures, causes and effects, and motion requiring a mover. What replaced it was a world of patterns, the laws of physics. After some important contributions from people such as René Descartes and Galileo Galilei, the first full-blown system of physical laws was introduced by Isaac Newton in 1687, the theory known as **classical mechanics**.

Pedantic but important aside: modern physicists distinguish between "classical" mechanics, which is a broad framework, and "Newtonian" mechanics, which is one specific model within that framework. Classical mechanics says that the world is made of things with definite, measurable values, obeying deterministic equations of motion; it stands in contrast with quantum mechanics. Newtonian mechanics adds specific ideas about absolute space and time. It stands in contrast with "relativistic" mechanics, which is classical but not Newtonian, and in which space and time become unified. Until we explicitly start talking about relativity, the equations we'll introduce for things like energy and momentum will be Newtonian. Just to make things one step more complicated, there are theories such as "Lagrangian mechanics" and "Hamiltonian mechanics," which are mathematically equivalent to Newtonian mechanics but feature different sets of words and concepts. Lagrangian and Hamiltonian mechanics are certainly classical; whether you want to characterize them as Newtonian is a matter of taste.

Classical mechanics is a theory of patterns, rather than natures or cause/effect relations, because it doesn't ask "What is the natural state of the system?" or "What caused the system to move in that way?" All it asks is "What is the system doing at this particular moment?" And from that it makes a precise prediction for what the system will be doing at any other moment. The moment could even be in the past, not just the future. Looking back at equation (1.2), conservation of momentum for two particles, the implication works backward in time just as well as forward. If we know the total final momentum, we know it was the same at earlier times.

This is an example of a different, much more grandiose, conservation law: **conservation of information**. This principle was implicit in Newton's laws of classical mechanics, but it wasn't brought out into the limelight until the work of French mathematician Pierre-Simon Laplace around 1814. The state of a classical system at any one time is specified by the positions and velocities of every part of the system, for example, all the planets in the solar system. That amount of information, Laplace pointed out, is preserved over time. From the state at any one moment, you can predict the state at any other moment, future or past. Or at least you could, if you had perfect knowledge of the information and arbitrarily accurate calculational abilities. Laplace imagined a "vast intellect" with such capabilities, but later commentators called this hypothetical being Laplace's Demon. The point of the Laplace's Demon thought experiment is not that anyone could actually have such information and make such predictions, in a practical sense, nor that we should aspire to do so. There's no way any realistic being could know the positions and velocities of every atom in a grain of sand, much less in the universe. But the universe itself possesses that information, and the laws of classical mechanics predict that it is conserved over time.

CONSERVATION OF ENERGY

Then we have **conservation of energy**, one of the most well-known features of classical mechanics and also an interesting example of the development of physical ideas. Unlike momentum, which is a vector, the energy of an object is simply a number, a quantity with magnitude but no direction. (Numbers are sometimes called "scalars" when we want to specifically distinguish them from vectors or more complicated quantities.) Energy comes in a number of forms, but there is one kind—**kinetic energy**, the energy of motion—that is related to momentum. The formula* for the kinetic energy of an object of mass m and speed v is

$$E_{\text{kinetic}} = \frac{1}{2}mv^2. \qquad (1.3)$$

Both momentum and energy are conserved in classical mechanics, but kinetic energy by itself is not, since it can be converted into (or created from) other kinds of energy. When you shoot an arrow from a bow, the energy that was tied up in the stretching of the bowstring is converted into the kinetic energy of the arrow.

In simple circumstances, we can directly track how energy transforms from one form to another. Physicists like to think about a ball rolling on a hill, where we imagine there is no friction or air resistance. Then there is **potential energy** associated with the elevation of the ball on the hill, in addition to the kinetic energy. The formula for potential energy of a ball at elevation h is

$$E_{\text{potential}} = mgh. \qquad (1.4)$$

---

* Why the ½? There is a reason, but it requires calculus to explain, and we haven't gotten there yet. For future reference: when we push on an object, the accumulated energy is the integral of the force as distance changes. Newton's second law equates force to mass times acceleration. Acceleration is the derivative of velocity with respect to time. So the integral of acceleration over distance equals the integral of velocity as the velocity itself changes. And $\int mv\,dv = \frac{1}{2}mv^2$. Perhaps none of that makes sense now, but just you wait.

Here, m is the mass of the ball, and g is the acceleration due to gravity near the surface of the Earth. (Feel free to substitute in the appropriate value for the acceleration if you're doing the experiment on some other planet.) Numerically, $g = 9.8$ (meters/second)/second—a dropped object (ignoring air resistance) increases its downward speed by 9.8 meters/second every second. That's what the acceleration would be if there were no hill at all.

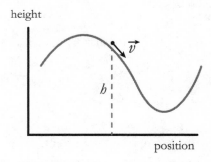

Rolling on a frictionless hill, the total energy $E_{kinetic} + E_{potential}$ remains constant, but the two forms of energy convert back and forth. If the ball starts motionless on a slope, it will start rolling downward, and at every moment its velocity will be exactly what is required to produce an amount of kinetic energy equal to whatever it has lost in potential energy.

It's easy to see potential and kinetic energy being converted back and forth into each other, but other forms of energy are harder to discern. We previously talked about the example of colliding billiard balls scattering off each other—in particular, "physicists' billiard balls," which move on a completely frictionless surface and create no sound or heat when they collide. In that case, both momentum and kinetic energy are conserved. At this point you may be getting flashbacks from a high school or college physics class. When both quantities are conserved, we are dealing with **elastic** collisions, since the objects involved simply bounce off of each other.

But there are also **inelastic** collisions, where momentum is conserved but kinetic energy transforms into some other form. Imagine that instead of billiard balls, we smashed together lumps of clay,
which each begin with an opposite amount of momentum,
$\vec{p}_1$ (initial) $= -\vec{p}_2$ (initial). Rather than ricocheting off, the two lumps
deform a bit and stick together. Momentum is still conserved, but kinetic energy by itself is not. It has been converted into heat and stress
inside the lump of clay.

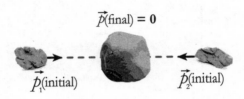

$$\vec{p}(\text{final}) = \mathbf{0}$$

$\vec{p}_1(\text{initial})$ $\vec{p}_2(\text{initial})$

Early thinkers, including Newton himself, didn't quite appreciate
that momentum and energy were two separate things; they thought
in terms of a single "quantity of motion." In the context of Newtonian mechanics, where constant straight-line motion for unforced
bodies is a fundamental principle, momentum is an obvious thing to
define. Kinetic energy was more subtle, although thinkers such as
Gottfried Wilhelm Leibniz (Newton's rival for the invention of calculus) argued that the central quantity in the study of motion was "vis
viva," which he defined as mv^2.

The situation was ultimately clarified by French philosopher and
physicist Émilie du Châtelet. She had translated Newton's work into
French and appreciated the conservation of momentum but argued
that energy was a distinct conserved quantity. As evidence, she
performed an experiment originated by Dutch physicist Willem
's Gravesande, in which heavy balls were dropped into soft clay. The
momentum of the ball is transferred to the Earth as a whole, but its
kinetic energy goes into creating a depression in the ground. The
amount of clay displaced was found to be proportional to the square

of the speed at impact, exactly what we expect from the formula for kinetic energy.

You may have heard that there is a law of "conservation of mass," but that hasn't been strictly true since the theory of relativity came on the scene. Under relativity, energy and momentum are both conserved (though the equations for them are slightly different from what we wrote above), but mass is just a particular kind of energy. That's the meaning of Einstein's famous equation: The energy of an object at rest (that is, with zero kinetic energy) is just its mass times the speed of light squared,

$$E_{rest} = mc^2. \tag{1.5}$$

In ordinary circumstances conservation of mass is a pretty good approximation, but in particle physics, where particles are routinely moving close to the speed of light, it's no longer useful, and we should just think of conservation of energy.*

WHY ARE THERE CONSERVATION LAWS?

Scientists delight in asking "Why?" questions. We want to know why apples fall from trees, why coffee and cream mix but never unmix, why a fire goes out if you deprive it of oxygen. But it's a tricky business. We have to face up to the possibility that our "Why?" questions bottom out with some ultimate answers, with nowhere further to go than "That's just the way it is."

Happily, when it comes to conservation laws we can do a bit better than that. It wasn't until the early twentieth century, but German

---

* You will also sometimes hear talk of "relativistic mass," which grows with velocity. That's confusing and unnecessary; it's better just to take the mass of an object as a fixed quantity and let the energy depend on velocity.

mathematician Emmy Noether* eventually proved a remarkable theorem relating conservation laws to **symmetries** of the laws of nature. The basic idea is straightforward: A symmetry is a transformation you can do to a system that leaves its essential features unchanged. A circle has a symmetry according to which it can be rotated by any angle around its center; a square, by contrast, is symmetric under rotations by 90 degrees or any multiple thereof.

Noether's theorem states that every smooth, continuous symmetry transformation of a system is associated with the conservation of some quantity. For example, the laws of physics overall are symmetric under shifts in space (we can pick the system up and move it somewhere else) and also under shifts in time (we can do an experiment, wait a little while, and do it again). In either case we should get the same answer before and after the shift. Noether's theorem relates these symmetries to conservation laws we already know: Invariance under spatial shifts leads to conservation of momentum, and invariance under temporal shifts leads to conservation of energy. The number of different symmetries works out. There is one dimension of time, and correspondingly a single conserved quantity: energy. But there are three dimensions of space, and we can translate separately in any one of those three directions. That's why momentum is a vector, which we can think of as having three components, one for each direction of space. There is also a symmetry according to which we can rotate the system around any of three independent axes; that gives rise to another conserved quantity: **angular momentum**.

---

* German vowels with an umlaut, such as ö, are often Anglicized to "oe." You will sometimes see the last name of Erwin Schrödinger, for example, written as "Schroedinger." You might therefore think that Emmy Noether's last name was really "Nöther," but that's not the case; she really did spell it "Noether." Johann Wolfgang von Goethe is a similar case.

Symmetry	Conserved Quantity
Temporal shifts	Energy
Spatial shifts	Momentum
Rotations	Angular Momentum

These symmetries—shifts in space, shifts in time, rotations—are all **spacetime symmetries**, since they involve transforming the system in space and/or time. In particle physics and quantum field theory we have **internal symmetries** that "rotate" different parts of a quantum field into one another. It's these internal symmetries that are responsible for conservation of electric charge and a number of other particle properties.

A subtlety arises when we remember that the laws of physics exhibit a certain symmetry, but that symmetry might be violated by the actual situation we find ourselves in. For example, our universe is expanding; faraway galaxies are gradually moving away from one another as time passes. Consequently, there is a sense in which energy is not conserved in an expanding universe. The configuration of our universe isn't invariant under time shifts; things used to be closer together, and in the future they'll be farther apart. If we simply take the energy contained in all the different forms of matter that we know about (radiation, ordinary matter, dark matter, dark energy, what have you) and add it all together, we get a number that is *not* constant over time. There are ways to try to fix this by defining an energy in the curvature of spacetime itself, but those ways aren't completely satisfying. There's nothing wrong with defining "the total energy contained in a region of space" as "the sum of the energies of all the things within that region," and admitting that this number changes over time.

All of which is to say: conservation laws are tricky. It's best to keep our wits about us. That's probably a lesson of broader applicability as we work our way deeper into the biggest ideas in the universe.

SPHERICAL-COW PHILOSOPHY

Conservation laws are obviously both conceptually important and practically useful. But there is another reason why they, and conservation of momentum in particular, are an appropriate starting point for an exploration of the physical world: They illustrate a basic methodological principle, the **spherical-cow philosophy**.

The name comes from a joke that physicists like to tell about themselves: A dairy farmer is struggling with milk output at the farm and decides to ask a scientist at the local university for help. For reasons that remain unexplained, they consult with a theoretical physicist. The physicist goes off to do some complicated calculations and returns with an impressive-looking stack of equations. "I've solved your problem, I think," says the physicist. "What is it?" replies the farmer excitedly. "Well, first assume a spherical cow . . ."

The humor value, in case it's not immediately obvious, lies in the fact that not only are cows not spherical but also that their non-spherical nature is crucial to what it means to be a cow. A spherical cow wouldn't be a cow at all. Making that assumption might simplify the calculations a physicist might want to do, but in doing so it removes us entirely from the realm of insights that would actually be useful to the dairy farmer.

The joke is famous not because it's side-splittingly funny—nobody ever claimed that—but because the equivalent of "assume a spherical cow" actually *does* work in physics, and it works incredibly well. It's an example of a general principle—namely, idealize a difficult problem down to a simple one by ignoring as many complications as you can. Get an answer to the simple problem. Then put the complications back in and calculate how they affect the answer to the simple problem.

This way of thinking is how we invented conservation of momentum. Aristotle wasn't wrong; coffee cups don't start moving by themselves, and if you push them for a while and then stop, they'll quickly come back to rest. But Ibn Sīnā wasn't wrong either; the cup comes to rest because of friction, not because of its intrinsic nature. If we can ignore friction and imagine an arrow flying through vacuum, it would continue moving at a constant velocity. *That's* the starting point for a useful physical analysis; we can always add the complications of friction back in later.

The master of this kind of reasoning was Galileo. He had a genius for figuring out what things were essential and what could be ignored at a first pass. Aristotle had claimed that heavier objects fell faster than lighter ones. Again, not wrong; drop a book and a single piece of paper at the same time, and see what your experimental results reveal. But Galileo argued that if there were no air resistance, they would fall at the same rate. That's an experiment no person could actually perform until centuries later, but Galileo was able to construct ingenious experimental setups that were able to establish the underlying principle.

We will see the spherical-cow philosophy at work again and again. But as useful as it is, it's not a universal truth-finding method. It doesn't work that well in dairy farming, for example. In many complex systems, such as we often encounter in the macroscopic human-scale world, different aspects of the underlying situation interact in crucial ways with one another, such that you can't just ignore them one at a time and correct for their effects later. In fields like biology or economics, everything often depends on everything else.

The reason why physics *seems* so hard is because it actually *is* quite easy, compared to other sciences. In physics (at least in some parts) we have this miraculous ability to set aside various components of the system we're thinking about, solve a much simpler problem, and put everything back together at the end. That makes our lives enormously easier than they otherwise would have been. As a result, physicists are able to discover amazing and counterintuitive features of the world, from quantum mechanics to relativity to the Big Bang. We would never have been able to just guess these things by being clever; we were forced to invent them by trying to fit the experimental data. But precisely because they are so counterintuitive, they can be hard to understand, at least at first glance. If the vista before us seems alien, it is because we are looking so very far.

TWO

CHANGE

Physics is made possible by the fact that the world exhibits a certain amount of continuity and predictability. But it would be a boring world indeed if all the things around us simply stayed the same from moment to moment. Happily, they don't. Planets and stars move through space, people commute to work or fly home for the holidays, individual atoms zip around inside your body and elsewhere. We need to turn our attention to understanding this behavior as best we can.

In classical physics, change is described using a specific framework we will call the **Laplacian paradigm**, after Pierre-Simon Laplace. The Laplace's Demon thought experiment illustrated conservation of information: The data needed to predict the future or retrodict the past is present at any moment in the history of an isolated system. This suggests the following paradigm for understanding how things change:

- Specify the complete state of the system at one moment in time.

- Use the laws of physics to calculate what the state will be just one moment later (or earlier).

- Use the laws of physics again, to evolve from that new moment to the next one.

- Repeat.

This algorithm allows us to build up the entire history of the system, past, present, and future. There is no reference to natures or purposes or goals toward which the system aspires. It's just one thing after another, where the next thing is determined by the current thing and the laws of physics.

But what in the world does it mean to say "just one moment later"? What is a "moment" supposed to denote here? A second? A millionth of a second? It seems a bit arbitrary.

Presumably what we want to do is to break time down into its smallest possible unit. But time seems to flow smoothly, whether from ordinary experience or from the perspective of modern physics. The continuity of time was recognized as puzzling even by the ancients, from Zeno of Elea (of "Zeno's paradoxes" fame) to Archimedes. The puzzle wasn't really resolved until Isaac Newton and Gottfried Wilhelm Leibniz independently developed **calculus**, the mathematical technique for dealing with infinitesimal quantities. So put aside any fears you may have acquired in high school or college; we're going to have to learn some calculus in this chapter.

The good news is that the basic concepts are easier to grasp than you might have been led to believe. There are only two of them, in fact: using "derivatives" to calculate the rate of change of something, and "integrals" to calculate the total amount of change.

(There is no bad news.)

PLANETS AND FORCES

To work our way into understanding the Laplacian paradigm for describing change, it's useful to contemplate the alternative. An example is provided by the motion of planets in the solar system, which has historically been a fruitful source of physical insight.

Most of us know the story of Ptolemy and Copernicus. Ptolemy, an Alexandrian astronomer of the second century, developed a geocentric (Earth-centered) model of the solar system that remained the state of the art for over a millennium. Copernicus, a Polish astronomer of the sixteenth century, put forward an alternative, heliocentric (sun-centered) model. This upset a number of people who preferred to think of themselves as living in the center of things. But both Ptolemy and Copernicus based their models on the geometry of circles, and as a result both needed to be wickedly complicated in order to fit the known astronomical observations. To model planetary orbits using nothing but circles, the orbits had to be individually offset rather than sharing a common center, and the planets didn't directly move on the main circles themselves, but rather on smaller circular epicycles centered on the main circles.

The situation was greatly simplified by seventeenth-century German astronomer Johannes Kepler. He accepted Copernicus's displacement of the Earth in favor of the sun at the center of the solar system, but Kepler's true outside-the-box insight was that, as awesome as circles undoubtedly are, there are other shapes we might contemplate for planetary orbits. He suggested that the orbits are **ellipses**, and in one fell swoop eliminated the need for epicycles and related complications entirely. On the basis of data that had been painstakingly collected by his mentor, Tycho Brahe, Kepler put forward three laws of planetary motion:

1. Planets move on ellipses with the sun at one focus of the ellipse.

2. The orbit of a planet sweeps out equal areas in equal times. Thus, planets move more quickly when they are closer to the sun, more slowly when they are farther away.

3. Larger orbits have a longer orbital period. In particular, the square of the period is proportional to the cube of the long axis of the ellipse. This relates the orbits of different planets to one another.

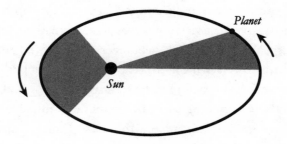

Kepler's laws represented an enormous step forward in our understanding of planetary dynamics. But as with any such advance, new questions instantly arose. *Why* do planetary orbits obey these laws? Is that even a sensible question to ask?

These and related questions preoccupied some of the best minds in the world at the end of the seventeenth century, the dawn of the Age of Reason. Inspired by work of Galileo and Descartes, natural philosophers trained their sights on the field of **mechanics**—the general study of motion and how it is produced. With the new understanding that momentum is conserved and straight-line constant motion was a natural state of affairs, it was important to understand why objects would ever deviate from straight lines. The obvious answer was that they were acted on by some kind of **force**, but at first it was somewhat obscure what counted as a force and how forces actually operated. It's not that different from Aristotle's concept of violent or unnatural motions, but with a different idea of what constitutes a natural motion—for Aristotle it's natural to be at rest, while in classical mechanics it's natural to move at any constant velocity.

An important step forward was made by Dutch physicist Christiaan Huygens. Imagine tying a rock to a string, then swinging it in a circle. You will feel the rock pulling on the string, and you have to pull back to keep it in circular motion. Huygens derived a formula for the amount of "centripetal force" you needed to exert to make this happen. (He also proposed the wave theory of light, invented the pendulum clock, and discovered Saturn's moon Titan. It was a heady time.) This isn't the same as planets moving around the sun, but the basic similarities were evident.

In England, the Royal Society was founded in 1660, and soon featured such thinkers as experimentalist Robert Hooke, architect Christopher Wren, and young astronomer Edmond Halley, all of whom became friends. Hooke had suggested in a lecture to the society that planetary motions could be explained by a gravitational force, stretching from the sun and decreasing with distance, that gently nudged the planetary trajectories away from straight lines. But Hooke wasn't fully comfortable with mathematics, and he wasn't able to build this idea into a comprehensive system. He would spend time in coffee shops (a popular innovation of the day) discussing the matter with Wren and Halley. By 1684, the friends had come to understand that Huygens's formula for centripetal force could be used to derive Kepler's third law, if the force of gravity from the sun obeyed an **inverse-square law**: The gravitational attraction decreases as the square of the distance to a planet. A planet that was three times as far away would feel one-ninth of the force. What they weren't able to prove was that an orbit under this kind of influence would actually take the form of an ellipse.

NEWTON AND HIS LAWS

Wren, who perhaps had more cash on hand than the others, from designing St. Paul's Cathedral and other buildings, offered a prize to anyone who could derive the shape of planetary orbits under such an inverse-square law. Even at that time everyone knew that Isaac Newton was the smartest one around, although he mostly kept to himself up in Cambridge. Eventually Halley worked up the courage to pay him a visit. Newton received him graciously, and Halley proposed the question of the shape of orbits subject to an inverse-square force. Newton instantly replied that it would be an ellipse. Somewhat taken aback, Halley probed further, asking how he could be so sure. "Why," Newton offered, "I have calculated it."

It remains a little unclear precisely how Newton carried out the original calculation. What we know is that Halley prodded him to write something up, which ultimately led to a short paper that Halley presented to the Royal Society. That first publication was lacking in some details, so Halley pressed the great man to flesh things out a bit. Once he set to doing so, Newton couldn't be stopped. Eighteen months later, he had produced *Philosophiae Naturalis Principia Mathematica* (*Mathematical Principles of Natural Philosophy*; usually shortened to "the *Principia*"), arguably the most influential work in modern intellectual history.

The *Principia* not only established the system of classical mechanics that remained unchallenged until the twentieth century. It also proposed the full law of gravitation, derived Kepler's laws from basic principles, and presented the first hints of what we now know as calculus. Newton had developed calculus pretty far, but he was reluctant to deploy it fully in the *Principia*, since the techniques were new and bound to be controversial. As a result, he ended up in a bitter dispute with Leibniz, who developed the ideas independently and whose notation eventually came into common use. (Newton also ended up in a bitter dispute with Hooke over the inverse-square law. He ended up in a lot of bitter disputes.)

Classical mechanics is a system, not merely a theory of some particular physical phenomenon. It's a comprehensive way of thinking about how the physical world works, one that was almost universally accepted as true until the advent of quantum mechanics. We've seen how a large number of very smart people struggled over the course of hundreds of years to get a grasp of momentum, force, and motion. Newton thought about it carefully and simply presented what seemed to be the right answer. As modern-day cosmologist Rocky Kolb has put it, "To compare the accomplishment of Newton to that of the first manned flight, one would have to imagine Orville and Wilbur Wright pulling up on the sands of Kitty Hawk on December 17, 1903, behind the controls of a modern jetliner and flying off to New York."*

So what did Newton actually say? We'll dig further into classical mechanics in the next chapter, but as far as the motion of the planets is concerned, there are two crucial ideas.

First, if the natural motion of objects is to continue in a straight line at a constant velocity, deviations from that motion can be characterized by the **acceleration** of that body, the rate at which the velocity changes. Like velocity itself, acceleration is going to be a vector, since we can imagine accelerating in different directions. **Newton's second law of motion**, the most famous equation in classical physics, states that the acceleration of an object is proportional to the net force acting on it, with the constant of proportionality provided by the object's mass:

$$\vec{F} = m\vec{a}. \tag{2.1}$$

---

* Rocky Kolb, *Blind Watchers of the Sky: The People and Ideas That Shaped Our View of the Universe* (Basic Books, 1997), 134.

The first law is that the velocity will be constant when there is no force, and the third law says that bodies act on each other with equal and opposite forces. Often there will be more than one force acting on an object at one time, in which case we simply add up the individual contributions to get the overall force, which is responsible for the acceleration. (Because vectors have a direction as well as a magnitude, two big forces can add up to give a small net force, if they're pointing in opposite directions or nearly so.) We can divide both sides of (2.1) by m to get the acceleration in terms of the force, $\vec{a} = \vec{F}/m$.

The second idea is **Newton's law of universal gravitation**, the final version of the inverse-square law that Hooke and others had talked about. What was indisputably Newton's was the idea that the law was indeed universal—it explains how apples fall from trees as well as how planets move around the sun. We take this for granted now, but back in the day it was a dramatic leap to connect planetary motion to everyday occurrences in the local orchard.

Consider two celestial objects with masses m_1 and m_2, and a distance r between them. Let $\vec{e}_r$ be a "unit vector"—a vector of length 1, in whatever units we are using to measure distance—pointing from object 2 to object 1. Then Newton says that the force acting on object 2, due to the gravitational pull of object 1, is given by

$$\vec{F} = G\frac{m_1 m_2}{r^2}\vec{e}_r. \tag{2.2}$$

The number G is a constant of nature, now known as **Newton's gravitational constant**, that tells us how strong the force of gravity is. From the point of view of object 2, object 1 is the **source** of gravity—the physical property creating the force—and vice versa.

This equation looks a bit more elaborate than our previous ones, but if we sit and think about it, we once again see a proportionality relationship between two vectors: the force $\vec{F}$ exerted on object 2 by object 1, and a unit-length vector $\vec{e}_r$ pointing from 2 to 1. The notational complexity simply reflects the fact that the proportionality factor involves Newton's constant as well as the two masses, and then we divide by the square of the distance. This attaches a precise numerical value to the amount of pull that the sun exerts on the planets, which is substantial when you are nearby and fades as you get farther away.

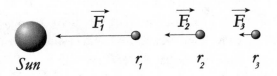

From these two simple rules, the second law of motion (2.1) and the law of universal gravitation (2.2), Newton was able to reproduce all of Kepler's laws, and much more. He showed, for example, that the gravitational force due to a spherical body is exactly the same as it would be if the same amount of mass were concentrated at a single point at the center. So it's okay to treat the sun and planets as if they were single points rather than solid objects, at least in the approximation where we treat them as perfectly spherical. Equally important, we can go beyond the idealization of treating each planet as orbiting the sun in isolation; with Newton's rules in hand, we can ask, for example, about how the orbit of Jupiter (the most massive planet in the solar system) affects the motion of the other planets. It was a whole new way of looking at celestial mechanics, which to this day is more than accurate enough to pilot a rocket ship to the moon.

THINK LOCALLY

Our aim here isn't to reproduce Newton's derivation of Kepler's laws, but to highlight the philosophical differences between two approaches to planetary dynamics (and therefore to physics more broadly). Kepler tells us that planets move in ellipses and says certain things about the speed at which they move. To describe an orbit as an ellipse is to make a **global** statement, in the sense that it refers to the trajectory as a whole. We wait for the planet to make one entire revolution around the sun, at which point we can verify that it really did move in an ellipse.

While Newton reaches the same conclusion, his procedure is very different. The starting point is now **local** in time: All of the relevant information is presented at a single moment. You tell me where the planet is, what its velocity is, and what forces are acting on it, all at some particular time. Then Newton's second law tells me the acceleration the planet is feeling at that same moment in time. From this, we can extrapolate the entire behavior at other times.

That's the Laplacian paradigm in action. Newton developed it, but Laplace emphasized its philosophical significance, and other researchers such as William Rowan Hamilton introduced significant upgrades to the idea. The Laplacian paradigm holds that all the information we need to determine what will happen to the system, or what did happen to it in the past, is contained in the **state** of the system at each moment in time. For a simple system like planets and the sun, the state is simply fixed by the positions and velocities of each component. You might protest that we also need to know the forces acting on the planets, and indeed we do; but those forces are determined by the positions and velocities of the *other* parts of the system. So if you have the positions and velocities of not just one planet but everything in the solar system, you are ready to go.

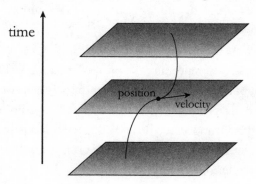

To complete this program, there are two big things we need to be able to do. First, we keep talking about the **rate of change** of various quantities. Velocity is the rate of change of position; acceleration is the rate of change of velocity. What do we mean by that, and how do we calculate it? If you take a journey from one location to another, you can figure out your *average* speed simply by dividing the total distance traveled by the time you spent traveling. But here we want the **instantaneous** velocity, and likewise the instantaneous acceleration, at each moment of time. That's not given by dividing any finite distance by any finite time. We'll have to be cleverer.

The other big thing is that, assuming we know the initial position, velocity, and acceleration, we want to build them up to construct the entire journey. In other words, we want to construct the **accumulated** distance traveled from the rate at which we were traveling at each moment. Again, in the simplifying case where we travel at a strictly constant velocity, it would be easy enough to simply multiply velocity by the time of travel to get the distance we traversed. But if we speed up or slow down along the way, or we turn to move in different directions, things are going to get trickier.

These two questions—what is the instantaneous rate of change of some quantity, and how do we accumulate quantities over time?—are precisely the subject of calculus. In math lingo they are called derivatives and integrals, respectively. Time to roll up our sleeves.

FUNCTIONS

Consider a car traveling along a straight-line road. Its position, velocity, and acceleration are all represented by vectors, but since there's only one direction of motion and therefore only one component of each vector, we can treat them as just numbers (that may be positive or negative). Let's further imagine that the car has an infinitely precise odometer, keeping track of exactly where its position is, and that we somehow record this information at every single moment of time.

Mathematically speaking, what we're in possession of is a **function**: the position of the car, x, as a function of time, t. A function is simply a **map** from one quantity to another quantity: I take some example of the first quantity, hand it over to the function, and it returns a value of the second quantity. The quantities in question might be single numbers, collections of numbers, or something more complicated. The input is called the **argument** of the function, and the output is the value corresponding to that argument.

$$\text{Function: } argument \rightarrow value$$
$$f\!: t \rightarrow x$$

A function that maps t to x might be written as $x = f(t)$, or just $x(t)$ for short. The actual letters we use to denote the functions and variables are completely up to us, as long as we remember what they're supposed to mean. If we specify positions in some region of land using coordinates x and y, we might denote the height of the terrain at each position by the function $h(x, y)$. So sometimes a quantity like x might be the argument (input) of a function; other times it might play the role of the value (output).

For our car, the argument is the time variable t, and the relevant function is $x(t)$, the value of the car's position at each time. We can represent it by plotting x as a function of t. You may be familiar with certain special functions, such as t^2 or $\sin(t)$, but the idea of a function applies whenever there is a unique map from values of t to values of x, regardless of whether there is a compact formula for expressing that map.

By "unique map," we mean that each value of the argument is associated with one and only one value of the function. That value may be repeated for different arguments—our car may pass through some point, back up, and pass through again—but at each t we better have a specific value of x. That means that when we plot a function, the curve may go up and down, but it will never double back from left to right.

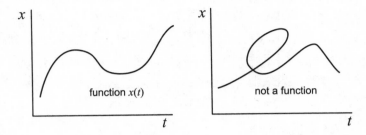

function $x(t)$

not a function

DERIVATIVES

Given $x(t)$, position as a function of time, we might want to ask what the velocity is at any moment. We know that the velocity is defined as the rate at which the position is changing, but how do we calculate it from $x(t)$? Velocity can't be determined from knowing the position at just one time—given the location of a car and no more information, we have no idea what its speed is. Somehow we need to take advantage of the information about the position at other times.

Just looking at the graph of the function, we get the general impression that the velocity of the car is related to the **slope** of the curve at each point, how sharply it is rising or falling. Where the curve is basically flat, time passes but the car doesn't move very much: The velocity is small. Where the curve is steep, the car moves a lot in just a little time: The velocity is large.

So we want to imagine drawing a straight line that exactly grazes our $x(t)$ curve at some time t_0, called the **tangent line** at that point. The velocity of the car at the precise moment t_0 is just the slope of that particular tangent line. We need to develop a systematic procedure for defining and calculating the slope of the tangent line at each point.

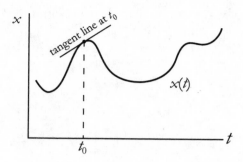

That would be easy to do if the car were moving at a constant velocity. In that case the function would just be a straight line, as in the next figure. Then the slope, and therefore the velocity, is straightforward to calculate: It's just the change in position divided by the corresponding change in time. In symbols, let's write the change in position as Δx and the change in time as Δt. Here Δ is the capital version of the Greek letter Delta, and is often used to represent the amount by which a quantity changes. (The notation Δx represents a single individual quantity, the change in x, not some quantity Δ times the quantity x.) Then the velocity along a straight-line path is

$$v = \frac{\Delta x}{\Delta t}. \tag{2.3}$$

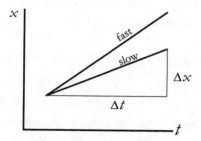

Now we come to one of the basic insights of calculus. As long as our function is relatively smooth—not overly jagged or randomly jumping from value to value—it will *look* straight over very short intervals, even if overall it's pretty curved. As we zoom in to higher and higher magnifications, a curved line begins to look straighter and straighter.

This suggests a strategy. Fix some time t at which we would like to calculate the velocity. Now pick some extra amount of time Δt, and consider the time interval in between an initial time t and a final time $t + \Delta t$. Given our function, we know both the position $x(t)$ of the car at the initial time, and also its position $x(t + \Delta t)$ at the final time. We can therefore calculate the total change in position over this time, which is simply

$$\Delta x = x(t + \Delta t) - x(t). \tag{2.4}$$

If the function as a whole is curved rather than straight, we can divide the total change in position by the total change in time, to find the average velocity over that interval.

$$v_{\text{average}} = \frac{\Delta x}{\Delta t}. \tag{2.5}$$

This looks similar to (2.3), but that expression was for the unique velocity along a constant-velocity path, while this is the average velocity for some specific time interval along any given path.

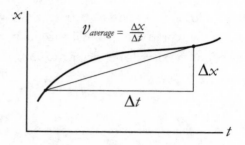

This isn't quite what we're looking for. We don't want the average velocity over some arbitrary interval, we want the instantaneous velocity any each moment. But you can maybe see where we're going. The time interval Δt that we chose to consider was arbitrary; we can make the interval anything we like. So let's zoom in. We can choose smaller and smaller values of Δt, implying smaller and smaller values of Δx. While both Δx and Δt can shrink all the way to zero, their ratio $\Delta x / \Delta t$ approaches some number that need not be zero. In fact, it approaches exactly what we're looking for—the slope of the tangent line at our initial point.

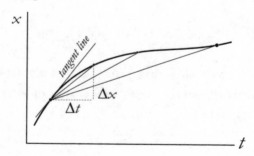

The procedure we've just outlined is called **taking the limit** as Δt approaches zero. Zero divided by zero isn't anything at all, mathematically speaking; it's just undefined nonsense. But we can take the limit of Δt and Δx as they both individually approach zero, and their ratio—the velocity, v—is well defined. We call this the **derivative** of the function $x(t)$, and write it in a suggestive notation as

$$v = \text{limit}\left(\frac{\Delta x}{\Delta t}\right) = \frac{dx}{dt}. \qquad (2.6)$$

That's it. That's what a derivative is: the slope of a curve at some point, defined by taking the limit of the slope of a sequence of lines that get closer and closer to the tangent line at that point. We considered the particular case of x as a function of t, where the derivative is the velocity, but the idea is much more general. The acceleration, for example, is the derivative of the velocity with respect to time,

$$a = \frac{dv}{dt}. \qquad (2.7)$$

Velocity is measured in meters per second, while acceleration is measured in (meters per second) per second, since it's the rate of change of velocity. The acceleration due to gravity of a freely falling object near the Earth is 9.8 meters per second per second (sometimes abbreviated as m/s^2).

Often we care about some function of x, which we can write $f(x)$, and its derivative will be df/dx. Every function is a function "of" some variable, and we can calculate the derivative with respect to that variable. It makes no difference whatsoever what symbol we use to denote the variable; that's just a label we choose at our convenience. It's useful to our comprehension to call time t and distance x, but that's totally up to us.

The quantities dx and dt are known as **infinitesimals**. They look like numbers we divide to get v, but the reality is a bit more subtle. If they actually were numbers, their value would be zero, and that's no good. Rather, they represent the idea of the limit of Δx and Δt as those two quantities both approach zero. The ratio of these two infinitesimals is a well-defined number, even if they individually are not. All I can say is, mathematicians have put an enormous amount of thought into making all this seem respectable. As physicists, we tend not to worry too much. If it works, we give the thumbs-up and move on to the next problem.

Two things might be bugging you at this point. First, it seems almost too easy. All we did was define the derivative by being very persnickety about the slope of a line tangent to a curve. Something as reputationally intimidating as calculus should be harder than that. And second, at the end of the story it's still not completely clear what we're supposed to *do* with this definition. It's a bit abstract. If we're handed an actual function, or the readout from an odometer, do we really have to go through all of this rigamarole to find its derivative?

These two issues are related, and they basically cancel each other out. If this were a real calculus class, you would immerse yourself in a tedious but reassuringly explicit set of rules for how to actually calculate the derivatives of given functions, a process known as **differentiation**. Consider a very simple function, $f(x) = ax + b$, where both a and b are fixed parameters (often just called constants).* This is known as a **linear** function, since plotting it would reveal a straight line.

We can find the derivative of this linear function simply by thinking about it. The constant b has no effect on the slope of the line, so we can ignore it. And the constant a just *is* the slope: If we change x by an amount Δx, then $f(x)$ changes by $a\Delta x$, so $\Delta f(x) / \Delta x = a$, no matter what x itself is. We therefore have

$$\frac{d}{dx}(ax + b) = a. \tag{2.8}$$

The derivative of our linear function is simply a constant, equal to the number multiplying x in the original function.

---

* Here x is the argument of the function, and $f(x)$ is its value. We're not assigning a separate variable name to the value, but you could if you wanted to, for example, $y = f(x)$. If we were to plot the function, x would be the horizontal axis, and $f(x)$ would be the vertical.

But differentiation is usually not so simple; most curves have a slope that changes from point to point. The derivative of the parabolic function $f(x) = x^2$, for example, is given by

$$\frac{d}{dx}x^2 = 2x. \tag{2.9}$$

We can see this in the figure, which plots the function $f(x) = x^2$ as well as the slope of the tangent lines at certain points ($x = -2, -1, 0, 1, 2$). For negative values of x, the parabola is sloping downward, and its derivative is negative. For positive x, the derivative becomes increasingly positive.

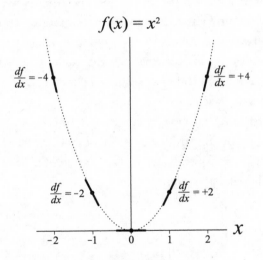

There are analogous formulas for the derivatives of x raised to other powers, and for the square root of x, and the logarithm of x, and sine or cosine of x, and the product of any two such functions, and so on. See Appendix A for some examples.

Learning all of these techniques is what makes calculus class a bit of a slog sometimes, but it's also what makes it useful to the working scientist. Our goal here is understanding the world as best we can, not training to become professional physicists. Basically we get to concentrate on just the fun parts. Differentiation gives us a way to calculate the slope of a curve, for example, to extract the velocity of a car from knowing its position over time. With that knowledge in hand, we can move on.

INTEGRALS

According to the Laplacian paradigm for classical physics, we can determine the future evolution of an object if we know its current position and its velocity, as well as those same quantities for all of the other relevant objects that might be affecting it. Then we can figure out what forces are acting on the object, and from Newton's second law (2.1) we can find its acceleration. With all that in hand, we'd like to actually figure out the object's trajectory. If velocity is the rate of change of position, and acceleration is the rate of change of velocity, we'd like to add up all those accumulated changes to determine how the position and velocity evolve over time.

Let's go back to a car moving in a straight line at constant velocity. But instead of plotting position as a function of time, let's pretend we don't know that, and plot velocity as a function of time. That's pretty easy when the velocity is constant.

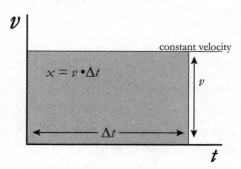

When something moves at constant velocity v, the distance it travels x is just that velocity times the time it spent moving, $x = v\Delta t$. There is a nice geometrical way of thinking about this in terms of the above plot: The distance traveled is just the area of the rectangle stretching horizontally from 0 to Δt and vertically from 0 to v. We can think of the accumulated amount of some quantity as the area under the curve defined by its function.

We'd like to be able to generalize this to the case when the velocity is not constant. When we were developing the idea of a derivative, the trick was to zoom in on a curve so that it looked like a straight line. We can use analogous reasoning here. The area under a curvy function $v(t)$ can be approximated by picking some small increment Δt, calculating the area under the rectangle from $t = 0$ to $t = \Delta t$ (which is just $v(0) \cdot \Delta t$), then doing the same for the curve from $t = \Delta t$ to $t = 2\Delta t$, and so on. At the end we add up all the areas of these skinny rectangles, and we get a quantity that is approximately equal to the area under the curve. The procedure of adding a lot of quantities together is represented by a summation sign, which is written as Σ, the capital Greek letter Sigma.

$$\text{Area under } v(t) \approx \sum\nolimits_{\text{rectangles}} v(t) \Delta t. \qquad (2.10)$$

The wavy equals sign is a way of writing "approximately equal to."

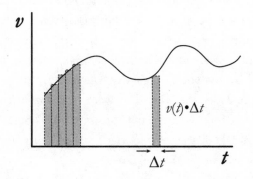

As our time interval Δt gets smaller and smaller and our rectangles get skinnier and skinnier (but also more numerous), this procedure comes closer and closer to the true area under the curve. We therefore, just as in the case of derivatives, take the limit as Δt gets infinitesimally small, and call it dt. We also replace the summation sign Σ with a new symbol that looks like an artistically deformed S. The result is the **integral** of the velocity with respect to time, which gives us the accumulated distance traveled:

$$\Delta x(t) = \text{limit}\left[\sum v(t)\Delta t\right] = \int v(t)dt. \qquad (2.11)$$

Similarly, the amount by which the velocity changes can be calculated as the integral of the acceleration:

$$\Delta v(t) = \int a(t)dt. \qquad (2.12)$$

We're hiding some information here to keep our notation clean. What we've calculated is the change in the position (or velocity) between some initial time and some final time; a more careful notation would indicate what those two times actually were. See Appendix A for details.

Just like derivatives can be thought of as a way of making sense of zero divided by zero, an integral can be thought of as a way of making sense of infinity times zero, where infinity is the number of skinny rectangles under the curve and zero is the area of each rectangle. It can all be made fairly rigorous (mathematicians refer to this subject as simply **analysis**). But you can see why Newton was hesitant to rely on calculus in the *Principia*. The ideas were very new at the time, and even today, some folks working in the foundations of mathematics are skeptical that calculus is *entirely* rigorous. Happily, it works more than well enough for physics purposes.

If we have Newton's laws and we know the forces acting on the object, we don't have to integrate anything to get the acceleration; it's given directly from $F = ma$. We can integrate acceleration to get velocity, and we can integrate velocity to get position. So we see the promise of the Laplacian paradigm fulfilled: Given an initial position and velocity, we are able to construct the entire trajectory of the object.

We can think of differentiation and integration as **operators** on functions: maps from functions to other functions. You give me one function, I can create another one by taking its derivative, or yet another one by taking its integral. In fact they are inverse operations of each other; the derivative undoes the integral, and vice versa.

$$\text{Derivative}\{\text{Integral}[f(x)]\} = f(x),$$
$$\text{Integral}\{\text{Derivative}[f(x)]\} = f(x). \qquad (2.13)$$

Or in symbols,

$$\frac{d}{dx} \int f(x)\, dx = f(x),$$
$$\int \frac{df}{dx}\, dx = f(x). \qquad (2.14)$$

As a matter of practice, calculating derivatives of functions is pretty easy, but taking their integrals can be quite difficult. Even the best physicists will often resort to using a computer to find a numerical solution when they need to know the value of a particular integral.

We won't worry too much about "doing" derivatives and integrals—taking some specified function and calculating the derivative or integral of it. Some examples are given in Appendix A, but mostly we are interested in the underlying concepts, rather than explicit calculation. We will use one simple result: The integral of the infinitesimal interval itself equals a finite interval:

$$\int dx = \Delta x. \qquad (2.15)$$

This equation simply says "The total amount of x we accumulate is the change in x from the start to the end of the process." The variable x here can stand for anything at all—distance in space, duration in time, whatever physical quantity we care about at the moment.

CONTINUITY AND INFINITY

Since the day Isaac Newton laid down his laws, a number of other kinds of laws have been suggested for fundamental physical systems. James Clerk Maxwell wrote down a set of equations for electricity and magnetism; Albert Einstein proposed an equation for the curvature of spacetime; Erwin Schrödinger suggested an equation for the wave function of a quantum-mechanical system; and many more. What all of these have in common is that they are **differential equations**—they involve derivatives (with respect to time, and often also with respect to space) of whatever substance is being described. That's why calculus is so central to how physics is done.

Does it have to be that way? We don't know the final laws of physics, so we should be open to different possibilities while we think about what they might be. One possibility, of course, is that the entire Laplacian paradigm is off base: that the fundamental laws are intrinsically global rather than local, so they can't be thought of as starting with some information at one time and chugging forward/backward to construct the entire solution.

Another possibility is that time is discrete, rather than continuous. Maybe there actually is a shortest possible time interval, and the universe evolves in units of this interval rather than smoothly. That's a pretty obvious possibility to contemplate, with some advantages and disadvantages. One advantage is that there are a number of mathematical and philosophical puzzles associated with the idea of continuity (and its close cousin, infinity), and perhaps the reason why they are puzzling is that they don't apply to the real world, so who cares? One disadvantage is that it would require us to toss out everything we thought we knew about classical mechanics and relativity, at least as part of the fundamental description of the world. That might ultimately be worth doing, but it's reasonable to approach such radical changes with a degree of caution.

To say that time (or some other quantity) is continuous implies that it takes on an infinite number of values, even between two points we think of as being a finite distance apart. You can convince yourself of this by imagining a line that represents time running from the past to the future. Pick out two moments, and label them $t = 0$ and $t = 1$.

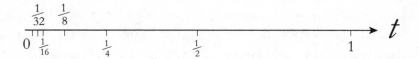

It's clear we can pick out a moment halfway in between, at time $t = \frac{1}{2}$. But then it's equally clear we can pick out a moment halfway between $t = 0$ and t $= \frac{1}{2}$, namely, $t = \frac{1}{4}$. And we can keep going with $t = \frac{1}{8}$, $\frac{1}{16}$, $\frac{1}{32}$ . . . And we didn't have to go between $t = 0$ and the next number, we could have gone between $t = 1$ and the previous number, obtaining $t = \frac{3}{4}$, $\frac{7}{8}$, $\frac{15}{16}$, $\frac{31}{32}$ . . . We can easily find an infinite number of numbers between any two points on a continuous line.

More remarkably, the infinity representing "the number of numbers between 0 and 1" is the same size as the infinity representing "the number of numbers between $-\infty$ and $+\infty$." That seems bizarre, since we tend to think that a proper subset should have fewer elements than the whole set. But infinity is special. We know that there are an equal number of numbers between 0 and 1 as there are between $-\infty$ and $+\infty$ because we can find an exact one-to-one correspondence between them. This correspondence can be represented by a simple function:

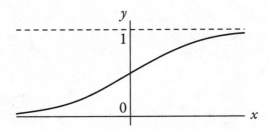

Here we have a function from x to y, with the property that any value of x between $-\infty$ and $+\infty$ gets mapped to some value of y between 0 and 1.

This might make you think that all infinite quantities are secretly the same size. Indeed, if you multiply infinity by 2 (or any other number greater than 0), you get back the same infinity. The number of integers is equal to the number of even integers. But things are not going to be so simple: There are different degrees of infinity, as discovered by German mathematician Georg Cantor in the late nineteenth century. **Cantor's theorem** shows there are infinitely many integers, and infinitely many real numbers, but the number of real numbers is greater than the number of integers. This discovery wasn't received with universal acclaim. Many mathematicians doubted Cantor's result, and his contemporary Leopold Kronecker called him a "corrupter of youth." Modern mathematicians generally accept the validity of Cantor's argument but continue to worry over whether we should trust the assumptions that go into it.

Does any of this matter for physics? Maybe not. We human beings are finite creatures, and one can take the stance that none of us can really discern a practical difference between "infinite" and "really big" (or "zero" and "really small"). So perhaps when it comes to describing the physical world, we can choose to be careful about infinity or not. But we shouldn't confuse what we human beings can hold in our heads with what nature actually does. It's conceivable that someday we will appreciate that when it comes to reality, there is one best way of thinking about issues of continuity and infinity, even if we don't know it yet.

THREE

DYNAMICS

Imagine two trees some distance apart, in a park where the ground is perfectly flat. Stand at one tree and point yourself in the direction of the other one. Now put on a blindfold and start walking. Let's assume you're really good at maintaining the direction in which you are moving, and that nobody plays tricks on you or gets in your way. At the end of your journey, you will find yourself at the other tree. And if you take off the blindfold and look back at your footprints, you will see that you moved in a straight line.

Now do something completely different. Take a long string, tie one end to one of the trees, and carry the other end to the other tree. Pull the string taut, so that there is the least amount of string we can use while still reaching between the two trees. The result is once again a straight line—the stretched string will lie exactly on top of the footprints from your previous journey.

This is pretty obvious and somewhat remarkable at the same time. We have the common-sense notion of "a straight line," but two distinct ways of constructing it. One is "keep moving in the same direction," and the other is "minimize the distance between your beginning and ending points." The first way reflects a local philosophy of action, in the spirit of the Laplacian paradigm we discussed in the previous chapter—at every moment of time you're doing a particular thing, and by the end all of your effort has constructed a certain path. The second way is more global, reminiscent of Kepler's laws—of all possible ways a string could stretch between the two trees, you're choosing the one that is the shortest. But these seemingly independent ideas are picking out the same answer at the end of the day.

Physics works the same way. We have emphasized, for good reason, Laplace's idea that information contained in the state of a system at a single moment can be used to construct its entire evolution through time, chugging implacably forward from one moment to the next. But there are other ways of getting the same answer, ultimately equivalent although very different in appearance. These other ways might invoke a rather different set of fundamental concepts in their formulation. This raises interesting questions about which vocabulary is "best" or "more real." If different ways of thinking are truly precisely equivalent, maybe that doesn't matter; but we don't know the ultimate laws of physics, and it's possible that one way of thinking will give us a more direct route to getting there. As Richard Feynman once said, two formulations of a theory might give exactly the same predictions and yet not be "psychologically identical when trying to move from that base into the unknown."

The last chapter focused on "change," whereas this one is about **dynamics**. The difference is that change is a completely general concept, whereas dynamics is specifically concerned with changes that obey the equations of physics. We'll be looking at the properties of specific physical systems and what classical mechanics tells us about their behavior. By thinking about kinetic and potential energy, we'll be led to interesting insights that relate the dynamics of different objects. This will ultimately inspire us to reformulate mechanics through a global view that considers the system's history as a whole—a remarkable idea known as the "principle of least action."

WHAT MATTERS TO MOTION

Let's think a bit more systematically about what's going on in the Laplacian paradigm. To keep things simple, we focus on a single particle moving in three-dimensional space, so its position is specified by a vector $\vec{x}$. The state of the system consists of both that position and the particle's velocity, which is the derivative of position with respect to time, $\vec{v} = d\vec{x}/dt$. Six numbers in total, for one particle in three dimensions: three components of position, and three components of velocity.

So the recipe is this. We have some specified situation, like "a ball rolling on a hill" or "a planet orbiting the sun." You give me the data $(\vec{x}, \vec{v})$ at some particular time t_0. From this, and our knowledge of the situation, we can calculate the total force acting on the object, $\vec{F}(\vec{x}, \vec{v})$. For example, if our object represents a planet, the force is given by Newton's law of gravitation applied to the sun and all the other planets. Then Newton's second law tells us the acceleration, $\vec{a} = \vec{F}/m$. Now we know not only the initial data $(\vec{x}, \vec{v})$ but also how rapidly they are changing:

$$\text{velocity } \vec{v} = \text{rate of change of } \vec{x} = \frac{d\vec{x}}{dt},$$

$$\text{acceleration } \vec{a} = \text{rate of change of } \vec{v} = \frac{d\vec{v}}{dt} = \frac{\vec{F}}{m}.$$

Then we can use calculus to move forward in time, constructing the entire trajectory $\left[\vec{x}(t), \vec{v}(t)\right]$.

This is a marvelously flexible framework. We've been talking a lot about particles, but classical mechanics is much more general than that. Let's say we want to describe an extended material, like a solid, liquid, or gas. And imagine we're treating it macroscopically, as a true continuum of matter rather than as a collection of individual atoms. What we can do is to zoom in on a tiny part of the object, an infinitesimal "volume element" dV of material. This element will be acted on by various forces; it will be tugged on by gravitational or electrical forces coming from the outside world, but it will also be pushed around by pressure from the other bits of the material that are right next to it. If we know the position and velocity of our little element, and the forces acting on it, Newton's laws tell us how it's going to move. After that, it's calculus to the rescue—we can derive equations for the system as a whole by adding up what happens to each little volume element.

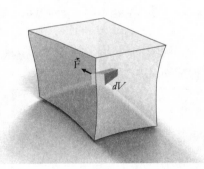

The fact that we need both position and velocity as input data to fix the trajectory of our system is important, but it's equally important that we don't need any *other* information about the initial motion of the system. We need to know what the acceleration is, but we don't need to be given it as an independent piece of information, since it's determined by Newton's second law (once we know what situation we're in and what the rest of the world is doing). Velocity is the derivative of position and acceleration is the derivative of velocity, so we say that acceleration is the *second derivative* of position:

$$\vec{a} = \frac{d\vec{v}}{dt} = \frac{d}{dt}\left(\frac{d\vec{x}}{dt}\right) = \frac{d^2\vec{x}}{dt^2}. \qquad (3.1)$$

Remember when you are looking at this notation that d is not a variable by itself; d/dt is an operator that takes the derivative with respect to time. The second derivative is when we operate d/dt twice on the same function, which is written d^2/dt^2.

We can also contemplate higher-order derivatives, which have been given whimsical names:

- Velocity = first derivative of position (with respect to time)

- Acceleration = derivative of velocity = second derivative of position

- Jerk = derivative of acceleration = third derivative of position

- Snap = derivative of jerk = fourth derivative of position

- Crackle = derivative of snap = fifth derivative of position

- Pop = derivative of crackle = sixth derivative of position

These terms are sometimes useful in engineering applications (and breakfast applications). But physicists rarely use them. For most purposes you simply don't need to—once you know the position and velocity of every piece of your system, you can figure out the accelerations and go from there.

GALILEAN RELATIVITY

Something quite profound is going on here. The fact that we need to be given the position and velocity of our system reflects the fact that there is **no preferred position or velocity** in the universe—no state picked out as somehow special, against which other states could be measured—at least as far as the laws of physics are concerned. For position this is not that hard to accept. If you do a physics experiment to test a fundamental law of nature in Florence, Italy, and then do the same experiment in Cambridge, England, you expect to get the same answer. You might get different answers for measurements of quantities that depend on your local environment, like the speed of sound (which depends on atmospheric pressure) or even the acceleration due to gravity (which depends on your elevation). But we don't expect the underlying laws themselves, like Newton's second law or the inverse-square law for gravity, to change from place to place. The laws of physics don't think of any locations in the universe as special.

The laws of physics also don't pick out any one velocity as special, although that's less intuitive. When you talk about the velocity of something, strictly speaking it is always measured with respect to something else. Given two objects, the distance between them is well defined, and their mutual velocity is the derivative of that distance with respect to time. There is no such thing as "the velocity," full stop. In our everyday lives this deep feature of reality is obscured to us, because we are never too far away from an obvious standard of reference: the Earth. When we talk about the speed of a car or plane, we are typically assuming that this speed is measured with respect to the Earth. But that's a feature of our local environment, not something built into the fundamental laws of physics. (And as pilots quickly learn, it's important to distinguish your speed relative to the ground from your speed relative to the air.)

Imagine you were sealed inside a spaceship with its engines off, not experiencing any acceleration. Without looking outside (or using instruments to do so), there would be no way for you to know how fast you were moving. That's because there's no such thing as "how fast you are moving," without specifying with respect to what. There is no absolute measure of rest in the universe, just as there is no preferred location.

The absence of a preferred standard of rest in the laws of physics was first pointed out by Galileo. He wasn't thinking about spaceships, but he did propose an equivalent scenario based on actual seagoing ships. His discussion was in the context of the then-current argument over whether the sun moved around the Earth or (as Galileo thought) the Earth rotated. Many people argued that if the Earth rotated, we would know about it, since the motion of the Earth's surface would be added to the motion of anything else. Galileo replied that all that mattered was the relative motion between two objects. For example, he posited that if you dropped a cannonball from the top of a ship's mast, it would drop straight down from the point of view of someone on the ship, whether the ship was stationary or moving with respect to the sea.

The emphasis on "relative" motion reminds us of the theory of relativity, and for good reason. The absence of a preferred location or standard of rest in the universe is known as the **principle of relativity**, and it was discussed by Galileo long before Einstein came on the scene. Newtonian mechanics is built on the foundation of **Galilean relativity**, which allows for any relative speed at all between two objects. The modern theory of relativity replaces that idea with

"Lorentzian" relativity, after Dutch physicist Hendrik Antoon Lorentz. The difference is that Lorentzian relativity includes an upper limit on the relative speed between two objects, given by the speed of light.

Notice that we've disclaimed the existence of a preferred position or a preferred velocity but haven't denied the existence of a preferred acceleration. That's because there is a preferred acceleration: zero. There is a special class of paths, known as **inertial trajectories**, which are those that aren't undergoing any acceleration at all. Unlike position or velocity, if you were sealed in a spaceship you could tell whether you were being accelerated; if you were, the spaceship would push on you in that direction.

BALL ON A HILL

In accordance with the spherical-cow philosophy, one of the most useful things we can do in physics is to pinpoint a simple idealized problem that is related to a large number of more realistic situations and understand that "toy model" (as physicists like to call them) as well as we can. We all have intuition about how the world works, but that intuition can be developed and extended by gaining familiarity with understandable examples.

There is no more common and useful toy model in physics than a ball rolling around in a hilly landscape. We already used it back in Chapter 1, when we talked about kinetic and potential energy. This example will turn out to be far more useful than you might initially have guessed; the insight we gain from thinking about balls rolling down hills will be directly applicable to quantum fields and the Standard Model of particle physics.

What we have in mind really is an idealized toy model, despite the fact that you have probably come into contact with actual balls rolling down actual hills at some point in your life. For one thing, as usual, we will completely ignore air resistance and friction and other ways energy can be dissipated. They're not hard to include, but our starting point is always to simplify down to the essentials, then add complications in later. More subtly, a real rolling ball will have not only kinetic energy and potential energy but also energy from the rotation due to its rolling. We will ignore that as well. What we imagine is a perfect featureless particle, moving without dissipation in a potential-energy landscape, with the sum of potential and kinetic energy remaining perfectly conserved.

According to the rules of classical mechanics, we want to specify the position and the velocity of the ball, then calculate the net force acting on it, which will tell us the acceleration. We imagine that the ball always stays on the surface of the landscape, not flying above it or buried underneath. As a further simplification, we'll consider motion along only one dimension, call it x. The ball also moves in a second dimension, height, but that's fixed to be the height of the landscape, $h(x)$. (We will not permit our ball to get hang time by flying off the hill.)

The net force is the sum of two different forces: the gravitational force pulling down, and the force exerted by the ground itself. The force from the ground is called the **normal force**, not because it's especially ordinary but because it's always perpendicular to the slope of the hill, and "normal" is just a synonym for "perpendicular" (as is "orthogonal" in this context).

Consider the simplest case we can imagine, where the landscape is flat, so that the height $h(x)$ is the same for every value of x. The gravitational force $\vec{F}_g$ pulls down, and the normal force $\vec{F}_n$ is perpendicular to the ground, so in this case it must push directly up. We therefore know instantly that there is no net force to the right or left, since the net force is the sum of one force directly down and one directly up. Furthermore, we've said that the ball won't be digging into the ground or flying up into the air, so the net force must be exactly zero. In other words, the gravitational force and the normal force are equal and opposite, and they add up to give no force at all, so there is no acceleration on the ball. This will be true no matter what the initial velocity of the ball is; whatever it is doing to start, it will keep moving in the same direction at a constant speed.

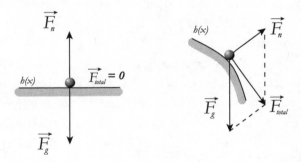

When the landscape is sloped, things get more interesting. Gravity still pulls the ball straight down, but now the normal force will not be vertical. As a result, when we add the two forces together, we'll have a nonzero net force. The total force, and thus the acceleration, will be in the direction of "down the hill." If we start with the ball at rest, it will start rolling downward; if it's moving down already, it will speed up; and if it's rolling up the hill, it will slow down.

Our rolling ball has both potential and kinetic energy. The potential energy is

$$V(x) = mgh(x), \qquad (3.2)$$

where m is the mass of the ball, g = 9.8 meters/second/second is the acceleration due to gravity, and $h(x)$ is the height of the hill at each point x. The notation here is a bit confusing if you're not used to implicit multiplication signs; m is a constant number, g is a constant number, and $h(x)$ is a function that yields a number h at each value of x. We multiply these three numbers together to get the potential energy, which is conventionally labeled V, although other letters are sometimes used. Since it's the energy that's going to matter to us, not the height of the landscape, we often speak directly in terms of **the potential** $V(x)$.

If we wanted to use Newton's second law $\vec{F} = m\vec{a}$ to solve for the motion of the ball, this potential-energy function gives us a nice way to think about it. The real ball moves both vertically and horizontally as the landscape undulates, so we have to think about adding vectors together to get the net force, as we discussed. But since we're imagining that the ball stays on the ground the entire time, it's really only horizontal motion that we need to calculate; the vertical ups and downs come along for the ride.

So we need to think about the acceleration in the x-direction, which comes from the force in the x-direction. Intuitively, we expect that force to come from the slope of the potential: The ball will be nudged down the slope, and the steeper the slope, the bigger the nudge. In calculus-speak, the force in the x-direction is minus the derivative of the potential with respect to position.

$$F_x = -\frac{dV}{dx}. \qquad (3.3)$$

The subscript on F_x is reminding us that it's the force in the x-direction. And the minus sign is just common sense: An upward-sloping potential (positive dV/dx) pushes the ball to the left, corresponding to a negative force in the x-direction.

THE ENERGY PERSPECTIVE

Once we know the force pushing on the ball, we could roll up our sleeves and solve for exactly how the particle will roll around in any specified landscape, using Newton's second law to calculate the acceleration and then doing some integrals. But that sounds kind of tedious. Let's instead get a bit of intuition for how things will go, just from thinking about energy conservation.

Along with the potential energy, we also have the kinetic energy,

$$E_{\text{kinetic}} = \frac{1}{2}mv^2, \tag{3.4}$$

where v is the speed of the ball. We add them together and get a number, the total energy, that stays constant over the ball's trajectory:

$$E_{\text{total}} = V + E_{\text{kinetic}}. \tag{3.5}$$

This simple equation by itself tells us quite a bit about the behavior of the ball. The kinetic energy is as small as it can be (namely, zero) when the velocity vanishes. So we instantly know that the potential energy, and therefore the height of the ball, is largest when the velocity is zero. Likewise, the velocity will be highest when the potential energy is lowest, at the bottom of the hill.

Conservation of energy gives us a nice global view of the ball's trajectory. Imagine we release the ball, with zero initial velocity, at the bottom of a valley. That's a minimum of the potential, where its derivative is zero. So there's no force on the ball, it just sits there. This accords pretty well with our intuition, that if we place a ball at rest at the bottom of a valley it won't move.

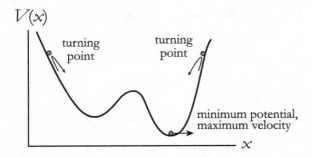

Now imagine starting the ball at rest again, but this time on a sloped portion of the hill. The total energy is simply the potential energy at this location. The ball will travel downward, speeding up and converting potential into kinetic energy. If the landscape starts sloping upward at some point, the ball will lose velocity as it now gains potential energy. And if the ball ever reaches the same height from which it began, its velocity will once again have to reach zero, since it will have regained exactly the potential energy it started with. That's known as a **turning point** for the trajectory, since the ball will then start rolling back in the direction from which it came.

In this idealized world of zero friction, the ball will travel forever back and forth between the two turning points. You might think it could reach the bottom and eventually come to a stop, but that's a case of our intuition misleading us: You're used to a world with friction. When the energy in the ball is precisely conserved, it oscillates back and forth for all eternity.

THE SIMPLE HARMONIC OSCILLATOR

There are spherical cows, and there are spherical cows. The favorite spherical cow in all of physics—the most important simple, exactly solvable physical system of amazingly wide-ranging applicability—is the **simple harmonic oscillator**.

Consider our ball on a frictionless hill, but now let's be a bit more specific about the shape of the landscape. In particular, let's make it a parabolic valley with a minimum at $x = 0$, so the potential energy is

$$V(x) = V_0 x^2. \tag{3.6}$$

Here V_0 is a parameter that tells us how wide (if V_0 is small) or skinny (if V_0 is large) the parabola is.

Just from looking at the potential we can easily figure out what the ball is going to do. If we start it with zero velocity somewhere on the right-hand side, say $x = x_0$, it will start rolling toward the bottom, then up the left-hand side. Since the potential is symmetric around $x = 0$, conservation of energy immediately tells us the ball will climb all the way to $x = -x_0$, where the potential energy will return to where it started. There the ball will once again have zero velocity and will start rolling back the other way. It will go all the way back to $x = x_0$, where the cycle will start again, and it will repeat forever.

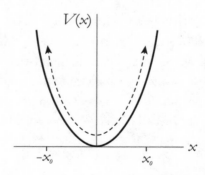

We could alternatively think in terms of forces rather than energies. From (3.3), the force in the x-direction is minus the derivative of the potential. If the potential is $V(x) = V_0 x^2$, and from (2.9) we know that $d(x^2)/dx = 2x$, then we have $F_x = -dV/dx = -2V_0x$. So when x is negative, the force is positive, pushing the particle back to the right; when x is positive, the force pushes it to the left. In either case, the force pushes the particle back toward the minimum at $x = 0$. This is called a **restoring force**, and its magnitude is proportional to the displacement $x = 0$ away from the equilibrium point.

This kind of system is called an "oscillator" due to its back-and-forth motion. It's "harmonic" when the potential is exactly proportional to x^2 (an x^4 potential would be an oscillator, but not a harmonic one), and "simple" when energy is exactly conserved (because there is no friction). There are also "damped" harmonic oscillators with non-zero friction, or "driven" oscillators where we put more and more energy into the system.

We can make a plot of how a simple harmonic oscillator behaves. Setting $x_0 = 1$ for simplicity, the particle starts at $x = 1$, moves down to −1, oscillates back to 1, and continues on in this pattern.

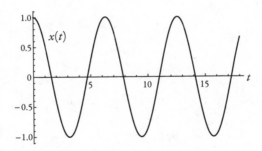

Functions of this type may be familiar (or cause flashbacks) if you've ever learned a bit of trigonometry. There are two especially important trigonometric functions: **sine**, which starts at zero, goes up to +1, then down to −1, then returns; and **cosine**, which starts at +1, goes down through zero to −1, then back again.

The easiest way to define these trigonometric functions is by thinking of a unit circle—that is, a circle of radius 1. Any point on the circle is uniquely specified by an angle θ from the x-axis. We're going to measure that angle in **radians**, units in which 360 degrees correspond to 2π radians, where $\pi = 3.14159\ldots$ is the famous numerical constant equal to the circumference of a circle divided by its diameter. (So one radian $= 180/\pi$ degrees.) There are a number of reasons for that, the most important of which is that the derivatives and integrals of sines and cosines take on elegant forms when we measure angles using radians. In these units, $\cos\theta$ is the projection of the point down to the x-axis, and $\sin\theta$ is the projection over to the y-axis.

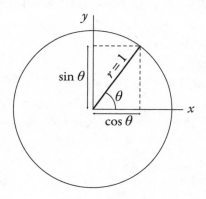

We can see that $\cos(0) = 1$ and $\sin(0) = 0$. From there, both functions oscillate up and down as the angle θ goes from 0 to 2π radians.

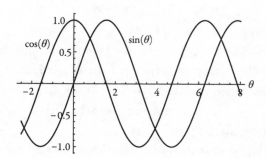

Comparing the plot of these trig functions to the evolution of the simple harmonic oscillator, we see that the oscillator's position looks very much like the cosine function. That turns out to be precisely true. For a general oscillator starting at rest, the position evolves as

$$x(t) = x_0 \cos(\omega t). \qquad (3.7)$$

The symbol ω is the Greek letter omega, which stands for the **angular frequency** of the oscillator. Whenever we have some oscillatory phenomenon, the frequency f is the rate at which the oscillator returns to its starting point, while the angular frequency ω is the rate at which the angle goes from 0 to 2π; they are related by $\omega = 2\pi f$. For the harmonic oscillator potential (3.6), the angular frequency is equal to $\omega = \sqrt{2V_0}$.

Now let's think about the velocity of the oscillator. It starts at zero, since we let go of our particle initially at rest. The particle starts moving to the left, so the velocity becomes negative. At the turning point the velocity returns to zero, and then it oscillates back and forth. This sounds somewhat like the sine function, except upside-down (since $\sin\theta$ starts at zero and goes up, while $v(t)$ starts at zero and goes down). That is also exactly right:

$$v(t) = -v_0 \sin(\omega t). \qquad (3.8)$$

The velocity oscillates with the same angular frequency as the position. The coefficient v_0 will depend on the mass of the particle represented by the oscillator; you could relate it to x_0 and V_0 by thinking about energy conservation.

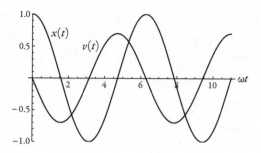

HARMONIC OSCILLATORS EVERYWHERE

Being able to solve equations of motion exactly, as we can for the simple harmonic oscillator, is satisfying as well as useful. It gives us a sense of accomplishment. And that ability is pretty rare in realistic physical situations. Even an unassuming fourth-power potential like $V(x) = V_0 x^4$ has no exact solutions expressible in terms of simple functions. The harmonic oscillator would be precious for that reason alone.

It's even better when your exactly solvable system appears over and over again in the real world. Happily, the simple harmonic oscillator does just that.

Think about a different kind of physical system, seemingly unrelated to a ball rolling on a hill. We consider a weight attached to a spring, suspended from above. If we pull the weight so that the spring stretches, the spring will exert a force pulling it back. But it's also true—at least in an idealized world where the spring is perfect and it doesn't bend or buckle—that if we push the weight upward, the spring will exert a force pushing it back down. There will be an **equilibrium** point in the middle where we let go of the weight and all the forces balance, so that it doesn't move at all. If we displace it a bit upward or downward from this equilibrium point, the weight will oscillate in an up/down cycle.

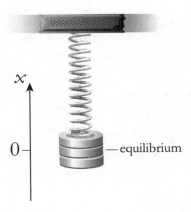

x

0 —   — equilibrium

It turns out that the vertical motion of our weight-on-a-spring system describes a sine wave, just like our ball in a parabolic potential. (Sines and cosines and any shifted versions thereof are often just called "sine waves," or "sinusoidal.") It's worth dwelling on this for a bit. The physical systems in these two cases are utterly separate: a ball rolling right to left in a valley, and a weight hanging from a spring. But the underlying *equations* are equivalent. From the abstract perspective of a theoretical physicist, they are the same system. (An experimentalist who has to build the system may disagree.)

There is a deeper reason why the simple harmonic oscillator is so ubiquitous in physics: An enormous number of systems, all the way up to vibrating quantum fields in the Standard Model of particle physics, are *approximately* simple harmonic oscillators. It's not hard to understand why.

Consider some physical system that oscillates back and forth with no friction, so that energy is conserved. There is some equilibrium point about which the system oscillates (or stays still, if the system starts there at rest). Let x be the quantity that measures how far the system is displaced from that equilibrium point. The potential energy is a function $V(x)$. For the moment we imagine that function to be completely arbitrary.

Now we deploy a crucial mathematical fact: We can express our potential-energy function as an **infinite series**, a sum of terms that each look like x raised to some power:

$$V(x) = a + bx + cx^2 + dx^3 + ex^4 + \ldots \qquad (3.9)$$

By carefully choosing the numerical coefficients $\{a, b, c \ldots\}$, any well-behaved function can be represented in this form. ("Well-behaved" excludes things like the function discontinuously jumping from one value to another.)

Let's think about this expression a bit. The first parameter, a, is just a constant number. It has no effect on the slope of the potential; different values of a correspond to moving the potential up or down without changing its shape. But it's only the slope of the potential that creates a force, not the potential's numerical value. From (3.3) we know that the force on the system is the derivative of the potential, which doesn't care about a. So we can just set $a = 0$ without changing the behavior of the system in any relevant way.

Now think about the rest of the infinite series. We have defined x so that $x = 0$ at the equilibrium point, where the system could just sit there without moving. What happens if we nudge the system a tiny bit? That corresponds to values of x that are very small. And when we take a small number (much less than one) and multiply it by itself, we get an even smaller number. So when we look at all the terms in (3.9), they become less and less important as we move further down the series to higher and higher powers of x. For sufficiently small x, all that should matter is the very first term, bx. This is an approximation, of course, but it's an approximation that is better and better as we consider smaller and smaller values of x. No matter what the other coefficients are, there will always be values of x that are so small that only the first term in the series matters.

Wait a minute. If $x = 0$ is the equilibrium point, that means $V(0)$

should be a minimum value—the bottom of the valley—where the slope is zero and there is no force. But in the small-x approximation where $V(x) \approx bx$, the slope of V at $x = 0$ is simply b. That's not zero unless b is itself zero. So given all of our assumptions, we can set $b = 0$ just as we set $a = 0$; otherwise $x = 0$ isn't a minimum at all. The potential we are left with is

$$V(x) = cx^2 + dx^3 + ex^4 + \dots \qquad (3.10)$$

And now we can make the same argument as before, that for sufficiently small values of x, the higher-order terms become irrelevant. In other words, on extremely general grounds we know that the shape of the potential for an oscillating system in the vicinity of equilibrium takes the approximate form

$$V(x) \approx cx^2. \qquad (3.11)$$

That's just a parabola, the potential for a simple harmonic oscillator. This is a wondrous result: For small deviations from equilibrium, in the absence of friction, *almost every oscillating system is approximated by a simple harmonic oscillator.** Our "oscillator" might be a ball rolling near the bottom of a potential, or a weight on a spring near its equilibrium point—but it could be something like the displacement of a pendulum, or an atom wiggling in a molecule, or the amplitude of a sound wave, or the electrical current in a circuit, or the value of the Higgs boson field. As long as it's a system that can be described in terms of kinetic and potential energy, the potential can be approximated near its minimum as a parabola, and therefore the physical

---

* Can you see why it has to be "almost" every? If we're unlucky we might find a system with $c = d = 0$, for example, when the potential is precisely $V(x) = ex^4$. Then there is no approximation in which the oscillator is harmonic, even for small x.

behavior is approximately that of a harmonic oscillator. The equivalence won't generally be exact; once we start caring about larger deviations from equilibrium, all those other terms in (3.10) will start to matter. But those are complications we can add in later.

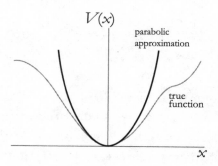

This discussion puts some meat on the bones of the spherical-cow philosophy. It's not just a matter of looking at oversimplified systems and hoping for the best. The idea of taking a complicated expression and writing it as the sum of infinite terms, as in (3.9), is a technique that is applicable to a wide variety of problems. And we are very often fortunate enough that the terms further along in the series are numerically smaller than the first few. This allows us to develop a systematic procedure known as **perturbation theory**: We write the equations governing the system as the sum of something simple plus a tiny perturbation, solve everything exactly for the simple part, then add in the rest bit by bit. Sometimes (not always) the universe helps us along in our attempts to understand it.

PHASE SPACE

According to the Laplacian paradigm, the trajectory of a system is determined by specifying the positions and velocities of each part of the system at some time. And we've said that the momentum of an object is just its mass times its velocity, $\vec{p} = m\vec{v}$. Since usually we consider systems for which the mass is constant, in terms of the information we need to predict the behavior of a system, specifying "position and momentum" is equivalent to specifying "position and velocity." Once we move to more advanced topics in physics—as soon as the next chapter—we'll see why momentum is more fundamental than velocity from a certain perspective, so we'll generally use that.

Together, the set of all possible positions and momenta for a system is known as that system's **phase space**.

$$\text{Phase space} = \{\vec{x}_i, \vec{p}_i\} \text{ for all objects } i.$$

(Curly braces {} are often used to denote a set of things.) Specifying where the system is in phase space at any moment is enough to fix its entire evolution according to Newtonian mechanics; in other words, *phase space is the set of all possible states the system could be in.*

As just a small hint to why momentum is more basic than velocity, think about Newton's second law, $\vec{F} = m\vec{a}$. We know that the acceleration is the derivative of velocity with respect to time. And this statement of Newton's principle implicitly assumes that the mass isn't changing with time, it's just a constant. So we can think of $m\vec{a}$ not only as "mass times the derivative of velocity" but equivalently as "the derivative of mass times velocity," since a constant can be brought inside or outside of the derivative:

$$m\frac{d}{dt}\vec{v} = \frac{d}{dt}(m\vec{v}) \text{ [when } m \text{ is constant]}. \tag{3.12}$$

This suggests an elegant way of rewriting Newton's second law in terms of the derivative of the momentum:

$$\vec{F} = \frac{d\vec{p}}{dt}. \tag{3.13}$$

Not only is this pleasingly compact, it's more general than $\vec{F} = m\vec{a}$, as this form remains valid even when the mass of the object is changing (for example, as a rocket gradually loses mass by ejecting exhaust). Force is the rate at which momentum is changing with time.

In the world of our experience, objects have locations in three-dimensional space. Mathematicians, and physicists following them, have repurposed the word "space" to mean something much more general—basically, any set with some additional structure. So "the set of all possible positions of a single object" is good old three-dimensional space. But the phase space for that object would be six-dimensional, three for the position and three for the momentum (which is a three-dimensional vector).

If our system consists of N particles in three-dimensional space, we can speak of the **configuration space** of the system, a $3N$-dimensional space labeled by the three-dimensional position of each particle. There will be a corresponding $6N$-dimensional phase space, since each particle will have a three-dimensional momentum as well. The Earth–moon system—two objects moving in three-dimensional space—has a twelve-dimensional phase space. Not to mention that if our objects are more complicated than simple particles, we might need to specify their orientations and angular momenta as well.

In general, phase space can be a very high-dimensional mathematical construction, depending on what kind of system we're thinking about. The idea was pioneered by Ludwig Boltzmann, James Clerk Maxwell, and other nineteenth-century pioneers of statistical mechanics who regularly contemplated systems with many particles. A traditional measure of "many" is **Avogadro's number**, 6×10^{23}, which is approximately the number of atoms in one gram of monatomic hydrogen. A gram is small but noticeable on human scales, so anything of that size or greater is safely macroscopic. A collection with Avogadro's number of particles is described by a 3.6×10^{24}-dimensional phase space. That's pretty big.

But phase space can also be pretty modest. The simple harmonic oscillator has a one-dimensional configuration space, labeled by its position x, so its phase space is two-dimensional, $\{x, p\}$. That makes it possible to visualize, and it's useful to do so.

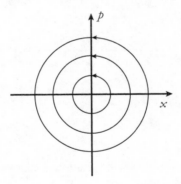

Here we have plotted three possible phase-space trajectories of a particular simple harmonic oscillator. The position oscillates back and forth, and so does the momentum, but with different starting points: When the position is at zero, the momentum is at its maximum, and vice versa. So any possible trajectory of the system describes an elliptical orbit in phase space, with ellipses of different sizes corresponding to different initial conditions. (It's an ellipse in phase space, not in real space; in real space the oscillator just goes back and forth. Phase space is a different idea. We don't need to independently specify the "velocity in phase space"—the entire trajectory is fixed by knowing what point you start at.) The frequency of the oscillator is independent of its initial amplitude, so the system takes the same amount of time to go around one circle no matter how big that circle may be.

Just for fun, let's ask what would happen if we added some friction into the system, to give ourselves a damped harmonic oscillator rather than a simple one. We intuitively know what will happen: The oscillator will rock back and forth, but it will lose energy, so the rocking will steadily diminish in amplitude. In the phase-space diagram, this corresponds to replacing our ellipses with spirals that move toward the center.

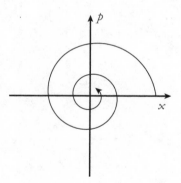

THE SPACE OF PATHS

That was a good amount of detail about balls rolling on hills and oscillators bouncing back and forth. Let's reward ourselves with something mind-blowing.

In exploring classical mechanics, we've emphasized the Laplacian paradigm, according to which we specify the state of the system—a point in phase space, giving the position and momentum of every part of the system at one moment in time—and that determines its entire trajectory. Harkening back to the opening of this chapter, this is much like putting on a blindfold and walking in a straight line. We know what we're doing at this moment, and we can chug forward in time from one moment to the next. Physicists think of this kind of procedure as an **initial-value problem**, since we start with some initial conditions and work out everything else from there.

But there are other ways of constructing a straight line, such as pulling a string taut between two trees. To make that line, we didn't have to think about "the initial direction" at all. We just needed the two trees, and pulling the rope between them automatically gave us a straight line pointing in the right direction. That's not an initial-value problem; it's a different kind of boundary-value problem, with one piece of information at the start (the location of the first tree) and another piece at the end (the other tree).

We can describe the string stretched between two trees in terms of global properties rather than local ones. (Recall the distinction between Kepler's theory of planetary orbits and Newton's.) "Walking in the same direction" is a quintessentially local procedure: You do something specific from moment to moment. Whereas "having the shortest length" is a global property: you need to compare the length of the entire string to what it would have been had it not been straight.

What we are implicitly doing is imagining an enormously big mathematical space, the space of all possible ways that a string could extend from one point to another one. Most of these ways will not be straight at all; there is a unique straight path, and an infinite number of curving ones. Within that gigantic set of possibilities, the straight path is distinguished by having a smaller length than any of the others.

Remarkably, all of classical mechanics can be cast in this kind of global language, rather than the local chug-forward-in-time perspective that we've been adopting thus far. Instead of specifying the position and momentum of a particle at some single moment of time, imagine that we specify just the position x_1 at some initial time t_1, and also the position x_2 at some later time t_2. Just like every curve in space has a length, every conceivable particle trajectory between (x_1, t_1) and (x_2, t_2) has (as we'll see) a quantity called the **action**, which depends on how the kinetic and potential energy evolve along the path. Out of all the possible trajectories the particle could take, the one it actually does take—the trajectory that obeys Newton's laws—turns out to be the one with the smallest action.

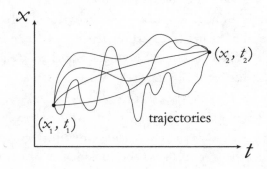

This idea is known, sensibly enough, as the **principle of least action**. This principle was gradually developed over a period of time from the 1600s to the 1800s. The principle of least action represents a departure from how we've been thinking, so let's pause to think about the mathematical philosophy behind it.

A large amount of math can be thought of as the study of *spaces and the maps between them*. A simple function $f(x)$, after all, is a map from the set of real numbers to itself. We've already been implicitly using many other examples. When we say "the position of a particle," we typically visualize a point literally embedded in space. But a mathematician would conceptualize that as a map from a single point to three-dimensional space.

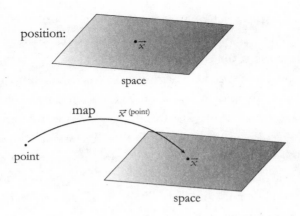

position:

space

map $\vec{x}$ (point)

point

space

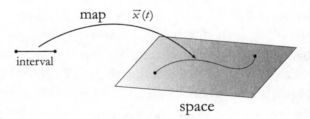

Similarly, the trajectory of a particle in space through time can be thought of as a map from an interval into space.

map $\vec{x}(t)$

interval

space

And for that matter, the notion of the length of a curve can be thought of as providing a map from the space of all curves (let's say, between two specified points) to the non-negative real numbers, where each curve is associated with its length L.

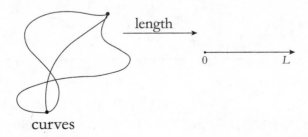

length

0   L

curves

Admittedly, "the space of all curves between two specified points" can be hard to visualize. We usually draw a few representative examples and hope that you can imagine the rest.

A point is a zero-dimensional space, a line is one-dimensional, a plane is two-dimensional, and so on. There is one dimension for each piece of information you need to specify where you are in the space. The space of all curves is therefore *infinite*-dimensional, since specifying a curve is to specify an infinite number of points along it. This creates some interesting mathematical subtleties, but the more notable thing is how much of our ordinary mathematical toolbox works on infinite-dimensional spaces without too much modification.

In particular, there is a way of doing calculus on infinite-dimensional spaces, called the **calculus of variations**. We're not going to dig into that, as fun as it would be; life is short, and there are a lot of big ideas ahead of us. But I want to mention it because calculus plays a crucial role in this idea of "find the curve of minimum length."

Think back to a simple function of one variable, $f(x)$. The derivative df/dx gives us the slope of that function at each point. Just from looking at the graph, we notice an important property: At every point where the function is at a local maximum (the top of a hill) or minimum (the bottom of a valley), that derivative is exactly zero.

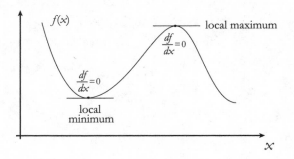

That's the mathematical secret behind finding curves of minimum length, or any other property. Consider the set of all possible curves.

We can define a function on that set (a map from each curve to some number) given by "the length of the curve." Then take the derivative of that function with respect to all possible ways we can infinitesimally deform the curve. If we're at the minimum, all those derivatives will precisely vanish. So we can turn the verbal idea "shortest-distance path" into a set of mathematical equations, "vanishing of the derivative of the length function on the space of paths."

LEAST ACTION

But it's not the length of a curve that we're interested in at the moment. When you throw a rock into the air and it follows some trajectory before hitting the ground, it's clear that the rock's path is not a straight line, which would be the shortest-distance path. What's minimized is not the length but the action, so we have to explain what that means.

At every point along its trajectory, a moving particle has a position x and velocity v, and correspondingly a kinetic energy $E_{\text{kinetic}} = \frac{1}{2}mv^2$ and a potential energy $V(x)$. We define the **Lagrangian** $L(x, v)$, after French mathematician Joseph-Louis Lagrange, as the kinetic energy minus the potential:

Lagrangian = kinetic energy − potential energy,

$$L(x, v) = \frac{1}{2}mv^2 - V(x). \qquad (3.14)$$

Then given any path $[x(t), v(t)]$, the action S of that path is the integral of the Lagrangian over time:

$$S = \int L(x, v)dt. \qquad (3.15)$$

The Lagrangian has some numerical value at every point along the path—that is, at each time. But the action is not a function of which point you are at on the path; it's a function of the trajectory as a whole.

The action for the actual path the particle will take, given its initial and final positions and times, will be smaller than the action would be for any other paths we could imagine between those points. This way of formulating classical mechanics is often known as **Lagrangian mechanics**, since the Lagrangian is the central quantity of interest. The action is the integral of the Lagrangian over time, and the real physical motions are those that minimize the action.

Let's see how the principle of least action works for a particle in a potential. It starts at position x_1 at time t_1, and ends up at position x_2 at time t_2.

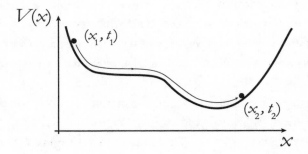

You might think: "I'll let the particle go at x_1 and calculate its acceleration using $F = ma$, then integrate up to get the velocity, until it reaches x_2." If so, you are still stuck in the Laplacian way of thinking. We can't just let the particle go with initially zero velocity, because it might reach the final point x_2 at the wrong time. Likewise, there's no requirement that the ball will *stop* at time t_2, just that it will be located at x_2 at that moment. The fact that we know the initial and final times as well as the positions is a strong constraint on what the particle can do in between. What the particle will actually do is take the trajectory that minimizes its action subject to those constraints, and that requirement will tell us what the initial velocity will have to be. Maybe it will have to roll up the hill at the start and away from x_2, just so it doesn't reach its final destination too early. The velocity will end up being whatever it needs to be to ensure that the particle ends up where it's supposed to at the appointed time.

What does it mean to minimize the action? The action is the integral of the Lagrangian, which is the kinetic energy minus potential energy. The kinetic energy $\frac{1}{2}mv^2$ is never a negative number, so to minimize the action we want to make the kinetic energy as small as we can. But we can't keep it exactly at zero—otherwise the particle wouldn't move, and in particular it wouldn't reach x_2 in time. It has to move fast enough to travel the right distance in the right time, but not too fast if it can help it.

Meanwhile, minimizing "minus the potential" seems easy enough—just travel up to somewhere where $V(x)$ is really high, so $-V(x)$ is really low. (For these purposes, a large negative number counts as "low.") But that comes into tension with the previous paragraph. If we start with the particle at x_1 and it zips off to somewhere the potential is much larger, it will have to quickly zip back to reach x_2 at the right time, and that would contribute a large kinetic energy. We're trying to keep the integrated kinetic energy as small as we can.

Minimizing the action sets up a balancing act: The potential energy wants to be high, but the kinetic energy wants to be low, and both have to compromise in order to begin and end at the specified points at the specified times. As a result, the trajectory that minimizes the action turns out to be precisely that which satisfies the classical equations of motion. The least-action way of doing things is mathematically equivalent to Newton's original formulation.

Which is remarkable, when you think about it. Newton's second law, the principle at the heart of classical mechanics, sets the force acting on an object equal to its mass times its acceleration. In the least-action approach, the word "force" doesn't even appear anywhere. It's the same dynamics at the end of the day, but a completely separate set of concepts is deployed to get us there.

So which is right? Does nature really start with some initial state and chug forward from moment to moment, as Laplace would have us believe? Or does nature have some kind of precognition, where it can visualize all the possible motions it might undertake between some initial point and some final point, and choose to move along the one that minimizes the action?

Neither one. Nature just is nature, and it does what it does. We human beings do our best to understand it on our own terms. It might turn out that we discover different equivalent ways to conceptualize the same underlying behavior. In those cases, it's less important to fret over which one is "right" than to be ready to think in whatever terms offer the most insight into the situation at hand.

FOUR

SPACE

So far we've been talking about things that happen—balls that roll down hills, weights that oscillate on springs, planets that orbit the sun. Implicit in all that discussion is the idea of an arena where these happenings take place. That arena is **space**, the collection of all possible locations of things. We've referred to the idea of space repeatedly—there it was, in the very first paragraph of Chapter 1— without too much deliberation, since presumably you are somewhat familiar with the concept. Now it's time to dig deeper. What properties does space have, and why? Why do we need something called space in the first place?

Some of these questions might have satisfying answers; some might not. What we can do is think about the properties that space has, and how they relate to other features of the physical world. Is space a thing, or just a property of other things? What does it mean to say that space is three-dimensional? And what are the physical behaviors that make it useful to think in terms of space? These questions will lead us to investigate yet another formulation of classical mechanics— after Newtonian mechanics and Lagrangian mechanics, here we will examine a picture called Hamiltonian mechanics. Unlike the other formulations, Hamiltonian mechanics doesn't treat space as special right from the start. It's therefore a helpful context in which to think about the special properties of space.

SUBSTANCE VERSUS RELATION

In the early 1700s, at the dawn of classical mechanics, a great deal of thought was given to the question of what space "really is." One idea was that space is a **substance** itself, with a separate existence from the things within it. The world, in this view, consists of various kinds of objects, *and* the space in which they are embedded. Space acts as a container for everything inside it.

That seems pretty natural to us, but there is an alternative view: that space isn't a thing at all, it's just a way to repackage the fact that any two objects are characterized by a quantity called "the distance between them." From this perspective, space is fundamentally **relational**. Once you know how all the objects in the world are related by the distances between them, there isn't any extra thing called "space" in which they reside. Any talk of "space" is just a convenient way of describing those distance relations.

These two opposing views on the nature of space, "substantivalism" and "relationalism," were taken up in a spirited exchange of letters between Samuel Clarke in England and Gottfried Wilhelm Leibniz in Germany. You will remember Leibniz as the co-discoverer of calculus with, and fierce rival of, Isaac Newton. The two had a number of feisty exchanges, although Newton preferred to work through intermediaries such as Clarke.

The Leibniz–Clarke correspondence was instigated by Caroline of Ansbach, a noblewoman who had been raised in Prussia, where Leibniz was one of her tutors. She moved to England when her husband, George Augustus, became Prince of Wales; later she would become queen when he was crowned King George II. Caroline had an abiding interest in science and philosophy. She became a champion of variolation—a precursor to modern vaccination—after directing a study on its ability to combat smallpox.

After Caroline arrived at the British court to become princess, Leibniz wrote to her to warn against aspects of English philosophy that he considered theologically problematic, especially those of John Locke and of course Isaac Newton. Caroline—mischievously?—brought these criticisms to the attention of Clarke, a prominent Anglican cleric and a friend of Newton's. Clarke wrote to Leibniz to defend the Newtonian view that space was absolute, against Leibniz's relational perspective, which had been heavily influenced by Descartes. We know at the very least that Newton offered advice to Clarke on what to write, although some scholars maintain that the letters were essentially ghostwritten by Newton.

The Leibniz–Clarke correspondence was cut short by Leibniz's death in 1716, but the surviving letters stand as a major work in the early philosophy of science. They discuss not only the nature of space but also free will and the essence of God. Today, most physicists would come down on the side of Newton, treating space as a thing in itself, for a couple of reasons. First, the space in between objects is not empty; it is filled with fields of various sorts. Second, space (as part of spacetime) has a life of its own; Einstein showed that the geometry of space responds to energy and can change over time, as we will discuss in Chapter 8. But we don't have the final answer.

And thus, a question that at one time was considered to be absolutely central to the progress of physics gets filed away in a drawer, without quite being answered. Part of the evolution of science is not simply the progress associated with learning new things; there is also the process by which we decide which questions are important, and which can be ignored. Sometimes it's because the question truly isn't interesting, other times the time for answering it is just not right. An exciting modern idea is that space emerges out of quantum entanglement, rather than being a substance in its own right, so maybe there's a sense in which relationalism will triumph in the end.

DIMENSIONS

Assuming we have a more-or-less intuitive grasp on what is meant by "space," let's consider some of its properties. The most important property is its **dimensionality**. The space through which we move in our everyday lives is three-dimensional. But physicists often look at lower- or higher-dimensional spaces. They might consider one or two spatial dimensions as simplified toy models, or more than three spatial dimensions in string theory or other models of unification. There are also other abstract mathematical "spaces"—sets with some additional structure, like the configuration space or phase space of a system of many particles—that may have a completely different number of dimensions. In this chapter we're focusing on the good old space around us.

Imagine picking up two long, thin rods. Tie them together so that they are perpendicular (at right angles) to each other. Now pick up a third rod and tie it to the first two at the same intersection such that the new rod is perpendicular to both the ones you started with. Now pick up a fourth rod, and tie it at the same place, so that it is perpendicular to all of the original three.

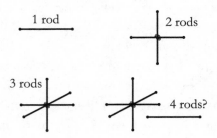

You can't do it. In our world, three lines can be mutually perpendicular to one another, but four cannot. That's one way of demonstrating that space is three-dimensional.

(You might know that a stool requires three legs to be stable. How many legs would be required if space were two-dimensional? Or four, or any other number?)

A dimension, then, is essentially "an independent direction in which things can move." In this conception, moving either one way or exactly reversed counts as the same direction. Reckoning from your present position, forward/backward is one direction, right/left is another one, and up/down is a third. All of these directions are independent, while other directions can be thought of as combinations of these. Three dimensions.

Another way of expressing the same idea is to note that three numbers, called **coordinates**, are required to specify the location of a point in space. Imagine picking some starting point and extending three abstract perpendicular directions, or **axes** (plural of "axis"), like an imaginary version of our perpendicular rods. We can assign coordinates to any point—for example, (x, y, z)—by traveling from the origin to that point by a series of straight perpendicular line segments, then measuring the length of each segment.

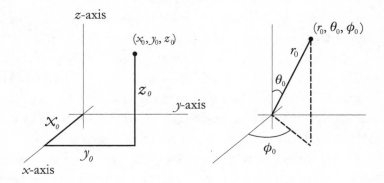

What's important is not the particular coordinate system that we use but the fact that it always takes three numbers to specify the location of a point. The set of perpendicular axes defines **Cartesian**

coordinates, named after René Descartes. But we could also specify the same point by measuring r, the distance from a radial segment from the origin to the point; θ (Greek "theta"), the polar angle between that radial segment and the z-axis; and φ (Greek "phi"), the azimuthal angle between the x-axis and the projection of our point down to the xy-plane. These are called **spherical coordinates**, since the set of all points at some fixed value of r defines a sphere.

Coordinates are not objective features of the world. They are human inventions, labels we attach to different locations in space. The rule of thumb is that no physical quantity should depend on what coordinates you use, just as the physical length of an object doesn't depend on whether it's expressed in centimeters or in inches. This property of **coordinate invariance** seems pretty straightforward, but souped-up versions of it lead us to contemplate symmetries and gauge theories, which play a central role in modern fundamental physics.

If space is three-dimensional, subsets of space can be lower-dimensional, at least to a good approximation. Consider a long, straight conducting wire. It's really three-dimensional, of course; if we look closely, we can see that the wire has a non-zero cross section. But when we look at it from far away, it appears essentially one-dimensional. If electrons moving along the wire tend to move rapidly and freely along its length, and only with great difficulty or not at all in the perpendicular directions, physicists will happily model that system as if it were one-dimensional. Likewise, there are cases when interesting phenomena are confined to two-dimensional spaces like thin films or the surface of a three-dimensional object; in that case we can often profitably treat the relevant phenomena as if they really lived in a two-dimensional world. This is the spherical-cow philosophy in action, approximating a complicated system as one that is so much simpler that entire dimensions disappear.

DIMENSIONS AND FORCES

Life would be quite different—indeed, biological life might not even be possible—if space really did have a different number of dimensions. Consider one of our favorite physical forces, gravity, described by Newton's inverse-square law: The gravitational force between two objects is inversely proportional to the square of the distance between them. Why is it the inverse of the square of the distance? Why not just the distance itself, or the distance to the eighteenth power?

A simple visualization is helpful here. Conjure in your mind's eye a set of straight lines stretching from the sun out to infinity in all directions. These are **lines of force**, so called because the gravitational force from the sun pulls objects along them. They're purely conceptual—there aren't really physical lines of anything stretching from the sun. But visualizing how these lines spread apart helps explain the inverse-square law.

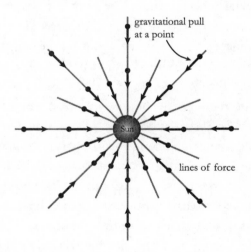

If we draw a sphere centered on the sun, the lines of force all pass through that sphere. If we draw another sphere with a bigger radius, the same lines will pass through it, but they will be more spread out—fewer lines will pass through any fixed area of the sphere.

The total area A of a sphere is related to its radius r by $A = 4\pi r^2$. The 4π in a formula like this is something we usually just memorize, but the fact that the area is proportional to r^2 just comes down to **dimensional analysis**. Physical quantities are always expressed in terms of **units**—length, time, mass, or some combination thereof.* Whenever we add things together, or equate them to one another, they need to have the same units.

Area always has units of "distance squared" (because it's a two-dimensional quantity), and the only distance involved in the question we're asking (what is the area of a sphere of radius r?) is the radius. So of course area is going to be proportional to r^2; there's nothing else it could be. Likewise, its circumference is measured in units of distance, so it will be proportional to r (it's $2\pi r$).

Gravitational lines of force don't begin or end in empty space; they originate at massive objects, which are the sources of gravity. So as we consider the lines emanating from the sun, it's the same total number passing through each sphere at different radii. The area of those spheres goes up as r^2. As a result, lines that were densely packed near the sun become more spread out as we move farther away. In particular, the density of lines goes down as $1/r^2$. And the number of lines passing through some object is proportional to the strength of the gravitational force.

That's the inverse-square law, right there. The force felt by an object is simply proportional to the density of force lines passing near it, which decreases as the reciprocal of the distance squared, because the lines are spread across a sphere whose area goes up as the distance

---

* Units are sometimes called dimensions, hence the name "dimensional analysis." But we'll reserve the word "dimensions" for the number of independent directions in which objects can move.

squared. (The same logic explains how objects look dimmer when we are far away, if we simply replace "lines of force" with "rays of light.")

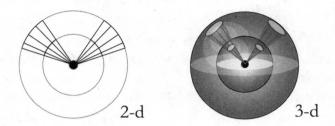

2-d   3-d

Things would be very different if space had a different number of dimensions. In two-dimensional space, an object would be enclosed by a one-dimensional circle rather than a two-dimensional sphere. So the density of force lines, and therefore the strength of the gravitational pull, would diminish as just the inverse distance, rather than the inverse distance squared. On the other hand, in four-dimensional space an object would be enclosed by a three-dimensional hypersphere. (This is the point where it's hard to visualize things, and we have to rely directly on the mathematics.) So gravity would obey an inverse-cube law. In general, in a universe where space is d-dimensional, the force of gravity is proportional to $1/r^{d-1}$.

Thought of this way, it seems almost inevitable that forces of nature should obey inverse-square laws. But they don't—at least, not all the time. At the level of elementary particles, there is a "strong nuclear force" and a "weak nuclear force" that reach over very short ranges before quickly diminishing to almost nothing. The reasons are different in each case. For the strong nuclear force, lines of force become tangled up with one another rather than stretching to infinity; for the weak nuclear force, the lines seem to gradually attenuate, but it's really because they are being absorbed by a Higgs field pervading all of space. Nobody ever said nature would be tidy.

There is one famous example, besides gravity, of an inverse-square

law: electromagnetism. The electric field of a charged particle gener-
ates lines of force that extend to infinity (unless they bump into other
charged particles, which often happens). As a result, the strength of
the electric force obeys an inverse-square law, in this case known as
Coulomb's law. At least in three-dimensional space.

The idea of lines of force was first suggested by Michael Faraday in
the mid-nineteenth century, in the context of electromagnetism. Faraday was born into a poor family, the son of a village blacksmith. As a
teenager he became the apprentice to a local bookseller, which gave
him the opportunity to educate himself through reading. He became
attached to the Royal Institution in London, first working on chem-
istry (he was the discoverer of benzene, as well as inventor of an early
version of the Bunsen burner) before becoming captivated by electricity and magnetism. His work in that field set the stage for James Clerk
Maxwell eventually bringing those phenomena together into a unified theory of electromagnetism. Maxwell had the mathematical
background to codify Faraday's physical insights into a system of rigorous equations.

MOMENTUM AND VELOCITY RECONSIDERED

What is it that makes space, "space"? That is, what are the properties
the world has that tempt us into describing it as "stuff, distributed
through space"? (And evolving through time, but we'll leave that
for the next chapter.)

To address this question, let's do one of our favorite things: introduce a new way of thinking about classical mechanics. We already
have the Newtonian way of doing classical mechanics, which is to
specify the positions and velocities of parts of a system and use Newton's laws to evolve them through time. And we have Lagrangian mechanics with the principle of least action, which tells us to imagine all
of the paths a system could take between specified initial and final

conditions, and pick out the one with the least action as the one the system actually takes.

Our third way of formulating classical mechanics is **Hamiltonian mechanics**. Its central idea is to elevate "momentum" to a concept with an existence of its own, independent of "velocity." This move can be a little tricky, because at first glance it seems almost indistinguishable from the Newtonian formalism, but it's different (and more powerful) in subtle ways. And those ways will be crucial for understanding why space is such an important concept.

In the last chapter we talked about phase space, the set of all possible positions and momenta that a system could have. According to the Laplacian paradigm, specifying one point in phase space at one moment in time is enough to determine the entire trajectory of a system (at least, one that is shielded from outside influences). And even though we used momentum as part of phase space, we know that within Newtonian mechanics momentum and velocity are interchangeable, since they are related by $\vec{p} = m\vec{v}$.

As successful as that picture is, there is a tiny bit of weirdness about it, so small that we didn't even remark on it at the time. Namely: the whole point of the Laplacian paradigm is that we tell you the state of the system at a single moment in time, and from there we determine the rest of the trajectory. But is "the velocity" really something defined at a single moment in time? Velocity is the derivative of position: the rate of change of position with respect to time. To calculate that derivative, we need to look beyond the single moment itself, to ask what the system is doing a moment later. Even if that moment is just infinitesimally later than the original one, it still seems like we are peeking just a bit at other moments in time, which fits awkwardly with the Laplacian philosophy.

HAMILTONIAN MECHANICS

The Hamiltonian approach resolves this awkwardness in an elegant way. Once again, we start with phase space, considered as the set of all positions and momenta a system could have. But now we insist that it really is the momentum vector we're talking about, not the velocity. In fact, from the Hamiltonian perspective, we *do not* define the momentum to be mass times velocity. The momentum is taken as a completely independent concept, equal in status to the position as the two things you need to specify to fix a point in phase space. A particle (or more complicated system) has a location in space, and it also comes with a vector property, "the momentum." This is true at any single moment in time, and we don't need to think about other moments in time, since for right now we're not thinking of the momentum as related to the rate of change of anything.

Here is how Hamiltonian dynamics works: Start with phase space, the set of all coordinates x and momenta p. (For simplicity we won't worry about using arrows over letters to denote vectors, or subscripts indicating which part of the system we're talking about; imagine that those are implicit.) Then we define a function, called the **Hamiltonian**, $H(x, p)$. The Hamiltonian is basically the energy of the system, written in terms of the positions and momenta.

We know that the potential energy of the system depends only on the positions, so we write it as $V(x)$. Then we have the kinetic energy, which in Newtonian mechanics we would write as $K = \frac{1}{2}mv^2$. But we'd like to write that in terms of the momentum, not the velocity. In Newtonian mechanics the momentum and velocity are related by $p = mv$. So we can get an expression for the kinetic energy in terms of momentum by substituting in $v = p/m$ to obtain $K = p^2/(2m)$. Altogether, then, we can write

Hamiltonian = kinetic energy + potential energy:

$$H(x, p) = \frac{p^2}{2m} + V(x). \tag{4.1}$$

So far, we've just done some manipulations to rewrite the energy in terms of position and momentum (rather than velocity). The cool new thing is that we can derive the equations of motion for the system from just this one expression. We'll explain how to do that in just a bit, but for now let's skip to the answers so we can get a bit of a payoff.

In Newtonian mechanics we track the evolution of just one variable, the position $x(t)$ as a function of time. Things like the velocity and acceleration are defined in terms of that variable (as the first and second derivative, respectively). There is correspondingly just one equation needed to figure out what happens, $F = ma$. But in Hamiltonian mechanics there are two variables, $x(t)$ and $p(t)$. So now we need two equations. And indeed we have them: equations for the derivatives of the momentum and the position, given by

$$\frac{dp}{dt} = -\frac{dV}{dx} \qquad (4.2)$$

and

$$\frac{dx}{dt} = \frac{p}{m}. \qquad (4.3)$$

The first of these equations is familiar—it's just Newton's second law $F = ma$ in different notation. Remember from (3.13) that we can replace ma with dp/dt, and from (3.3) we know that the force is minus the derivative of the potential with respect to position. The second equation, (4.3), is also familiar—it's just a rewrite of $p = mv$, momentum is mass times velocity. But the *interpretation* of this equation is different, and the difference is crucial. To grasp the distinction is to appreciate Hamiltonian mechanics in your bones.

In the Newtonian perspective, the fundamental object is $x(t)$, the path defined by position as a function of time. From that, everything follows; we can take a derivative to get the velocity, and we define momentum as mass times velocity. Momentum is not defined that way in

Hamiltonian mechanics; it is a variable in its own right. The relationship (4.3) looks the same in both approaches, but from the Hamiltonian perspective it is an **equation of motion**—a relationship that holds true on actual, physically realistic paths, not something true by definition.

What's the difference between a definition and an equation of motion? A definition implies an inevitability to the relation; in Newtonian mechanics, there is nothing the momentum could have been other than mass times velocity, since that's what it was defined to be. An equation of motion, by contrast, picks out the "right" relationship between variables, even though we can imagine them taking on other (wrong) values.

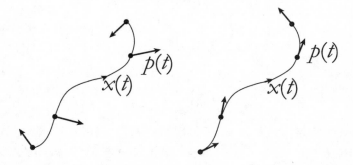

Two trajectories in phase space, by which we mean a point in space $x(t)$ and a momentum $p(t)$ for each time t. On the left, the momentum vectors are not proportional to the velocity—they can't be, since they don't point along the spatial path—so this trajectory does not obey the equations of motion, whereas the one on the right does.

In Hamiltonian mechanics, momentum has an existence independent of velocity. We are free to imagine any trajectories for the position and for the momentum, $x(t)$ and $p(t)$, including ones where p is not equal to *mv*. In the space of all conceivable behaviors, there is no necessary relationship between position and momentum. But there are certain special trajectories, those that obey the equations of motion. And on *those* trajectories, momentum equals mass times velocity.

If your brain insists on thinking of momentum as mass times velocity, imagine replacing "momentum" in the above discussion with some other label, like "purpleness." Then Hamiltonian mechanics says that the state of a system at any one time is defined by its position and its purpleness vector. On physical trajectories that obey the equations of motion, purpleness happens to equal mass times velocity, but we can imagine ones where it doesn't. It's somewhat inconvenient that we used the label "momentum" for both the independent variable in phase space in the Hamiltonian picture and for mass times velocity in the Newtonian picture; they're numerically equal when a trajectory obeys the equation of motion, but they're conceptually distinct.

HAMILTON'S EQUATIONS

That's the philosophy of Hamiltonian mechanics: Treat position and momentum as two conceptually distinct variables, and determine equations of motion for each of them. The "Hamiltonian" itself is simply the energy of the system, expressed as a function of position and momentum (and not as a function of velocity or other derivatives). The idea is that we should be able to use the explicit expression for the Hamiltonian, (4.1), to derive equations of motion for momentum (4.2) and position (4.3). How exactly does that work?

Let's think about the equation of motion for momentum, (4.2). It says that the rate of change of momentum with time is equal to (minus) the slope of the potential-energy function. This is morally equivalent to Newton's second law, $F = ma$, since the slope of the potential is (minus) the force on the object. We can think of momentum as being pushed around by the way the potential energy changes with position. That makes physical sense: If the potential were completely flat (zero slope, like a ball rolling on a featureless table), momentum would be conserved.

But the potential energy, we notice, is one of the two terms in the Hamiltonian. The other is the kinetic energy, $p^2 / 2m$. If momentum is pushed around by the slope of the potential energy, maybe there's a nice kind of symmetry where position is pushed around by the slope of the kinetic energy?

That's exactly what happens. Remember from back in equation (2.7) that the derivative of x^2 with respect to x is $2x$. That works no matter what variable we are considering, so, for example, the derivative of p^2 with respect to p is $2p$. The derivative of the kinetic energy with respect to the momentum is therefore

$$\frac{d}{dp}\left(\frac{p^2}{2m}\right) = \frac{1}{2m}\frac{d}{dp}\left(p^2\right)$$
$$= \frac{1}{2m}\left(2p\right)$$
$$= \frac{p}{m}. \tag{4.4}$$

(Whenever we take the derivative of a function, constants like $2m$ just go along for the ride; in this case, there are two factors of 2 that cancel each other out.)

Interesting! Just as we had speculated, taking the derivative of the kinetic energy with respect to momentum gives us the right-hand side of (4.3), the equation of motion for the position. There is an annoying minus sign in one of the equations and not the other, but roughly speaking the momentum is pushed around by changes in the potential energy, and the position is pushed around by changes in the kinetic energy. To summarize:

Rate of change of momentum with time = Minus the slope of
potential energy with respect to position;
Rate of change of position with time = Slope of kinetic energy
with respect to momentum.

Together, these two relations are known as **Hamilton's equations**.

It's a beautiful structure, but there's a technical point we need to think about a little more. To keep things concrete, we've been focusing on a single particle moving in a one-dimensional potential, a standard and illustrative toy model. In that case, the Hamiltonian is given by the sum of the kinetic and potential energies, as in (4.1).

But the formalism of Hamiltonian mechanics is much more general than that. The Hamiltonian of a system will always be a function of some set of coordinates and their corresponding momenta, but there is a lot of freedom in what the actual function is. We can have any number of positions and momenta; in field theory there will be an infinite number. For that matter we are allowed to consider quite complicated Hamiltonians, even ones where positions and momenta mix together so that it's not possible to distinguish between "potential energy" and "kinetic energy." Much of modern physics consists of guessing the right Hamiltonian for a system of interest; once you get that, you know how the system will behave. We need to think a bit more generally.

PARTIAL DERIVATIVES

What we want is a version of Hamilton's equations that works for any Hamiltonian whatsoever. And to get that, we're going to have to introduce one last bit of calculus: **partial derivatives**.

We need this new technology because we want to derive our equations of motion from a single expression, the total Hamiltonian, rather than the potential energy and kinetic energy separately. But unlike the potential energy, which depends only on x, or the kinetic energy, which depends only on p, the Hamiltonian depends on both x and p at the same time. We need to be able to take the derivative of a function of more than one variable.

The derivative of a function is just its slope at each point. But when the function depends on two or more variables, we don't simply have a curve with a unique slope at each point. It's more like a meandering landscape, where there is more than one direction we could move in. Some directions would take us upward, some downward, some would stay at the same elevation. We're going to have to be more nuanced than just defining "the slope" once and for all.

Partial derivatives provide a clever answer to this question. Consider a simple function of two variables, x and y, given by x times y^2 times a constant, a:

$$f(x, y) = axy^2. \tag{4.5}$$

The idea behind partial derivatives is that when we have a function of several variables, we can take turns with them. That is, we take derivatives with respect to each variable separately, and when we do that, we treat all the other variables as if they were constants. And to let people know that's what we're doing, we replace the d in our derivatives by a new symbol: ∂. (Sometimes called "del," but that word is also used for other symbols, so it's best just to call it "partial" when you're reading out loud.)

There are thus two different partial derivatives we can take of $f(x, y)$: the partial with respect to x, and the partial with respect to y. When we take the partial with respect to x, we differentiate with respect to x while treating both a and y as constants. This gives us

$$\frac{\partial}{\partial x}\left(axy^2\right) = ay^2 \frac{d}{dx}(x) = ay^2. \tag{4.6}$$

Whereas, when we take the partial with respect to y, we treat a and x as constants, and end up with

$$\frac{\partial}{\partial y}\left(axy^2\right) = ax\frac{d}{dy}\left(y^2\right) = 2axy. \qquad (4.7)$$

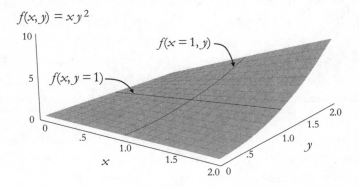

$f(x,y) = xy^2$

A function of two variables, $f(x, y) = xy^2$. We can take a partial derivative with respect to x by thinking of y as a constant and differentiating the resulting function; likewise we can take a partial derivative with respect to y by thinking of x as a constant.

That's basically it. To take a partial derivative, you specify which variable you're differentiating with respect to, and you treat all the others as if they were constants (and therefore unaffected by the derivative operation). The task of actually solving partial differential equations can be wickedly difficult—people devote their entire professional lives to coming up with better ways of doing it—but we don't care so much about that here. We just want to be able to see where Hamilton's equations of motion come from.

Now we can put it all together. We know that we can get equations of motion for momentum and position by taking the derivative of the potential and kinetic energy, respectively. And the Hamiltonian is the total energy, so (in this simple example) the sum of those two. Partial derivatives let us take derivatives of each variable separately. We can therefore present Hamilton's equations in their full generality:

$$\frac{dp}{dt} = -\frac{\partial H}{\partial x}, \ \frac{dx}{dt} = \frac{\partial H}{\partial p}. \tag{4.8}$$

I promise there are no typos. There really is a minus sign in the first equation but not the second one. There are reasons for this, but they are lurking deep within parts of mathematics that we won't be going into here. (Look up "symplectic geometry" if you're interested.) And there really are ordinary derivatives with d's on the left-hand side of each equation, and partial derivatives with ∂'s on the right-hand side. That's because position and momentum are functions of just one variable—time—whereas the Hamiltonian is a function of both position and momentum, so we need partial derivatives that specify what we're differentiating with respect to.

It's an elegant approach. In the Newtonian philosophy, each part of the system would have an equation of motion, describing which forces were acting on it. After the fact, we could show that certain quantities like "energy" were conserved. In the Hamiltonian philosophy, by contrast, we write down just one equation—an expression for the Hamiltonian, the total energy as a function of position and momentum. From that we can derive all the necessary equations of motion. This continues to be true when we have much more complicated systems, featuring multiple parts interacting in any number of ways. There will still be just one Hamiltonian describing the whole shebang; everything about the dynamics is implicit in that expression. Newtonian mechanics, Lagrangian mechanics, and Hamiltonian mechanics are equivalent as physical theories, but taking one approach or the other might make our lives easier in different circumstances.

I did say there would be no homework, but if you wanted to get your hands dirty you could work out the explicit form of Hamilton's equations for the simple harmonic oscillator, where the potential energy is $V(x) = \frac{1}{2}\omega^2 x^2$. Or even make up your own Hamiltonian— perhaps two oscillators, with some kind of interaction between them?—and see what happens.

LOCALITY

Who cares? Why are we putting ourselves through all of this mathematical effort just to derive a set of equations that, at the end of the day, just recapitulate Newtonian mechanics in a more intimidating notation?

There are many reasons to become comfortable with the Hamiltonian version of mechanics, not least of which is that it becomes indispensable when we make the transition to quantum mechanics. But for our present purposes, thinking in this way helps us see what is so special about space.

From the Newtonian perspective, "space"—that is, "position space," the set of all possible locations—is clearly special from the start. We can think about "momentum space," the set of possible momenta, but it seems a bit abstract, unlike the straightforward nature of position space. We live in position space, after all, not in momentum space.

But when we turn to the Hamiltonian perspective, positions and momenta seem to be on the same footing, at least to start. They are the two coordinates on phase space. The Hamiltonian itself is a function of both x and p. And when we look at Hamilton's equations (4.8), position and momentum are nearly symmetric (other than that minus sign). As far as the general structure of Hamiltonian mechanics is concerned, the Hamiltonian could be any function $H(x, p)$ at all. There's nothing in the formalism that distinguishes between position and momentum to indicate "this is the one where we live."

So in this setting we can ask a question that wouldn't even have occurred to Newton or his immediate successors: What's so special about space? Why do position and momentum seem so different to us in practice if they appear somewhat equally in the Hamiltonian laws of physics? Why do we live in position space, and not in momentum space?

The thing that makes space special is that interactions are **local** in position. Objects directly interact with one another when they are in the same location, not when they have the same momentum (or whatever). This might not be obvious—doesn't the gravitational force from the sun stretch across space to the planets?—so we have to think about it a bit.

As physicists, it's fun to talk about the behavior of a single system all by itself. But that's pretty meaningless if we can't also imagine observing the system and figuring out what it's doing. Ultimately we need to describe multiple systems interacting, influencing and being influenced by one another. And it's a feature of the universe in which we live that systems generally interact when they are next to one another in space. That's what physicists mean by "locality": Doing something at a particular point in space only affects other things at or next to that point. Eventually those effects can ripple out to other locations, but that takes time.

Think of billiard balls rolling around a table. They generally move on a straight line (unless we're doing a trick shot with significant spin on the ball) until they collide with the bumpers or another ball. And that happens when they come to the same location in space. It doesn't matter what their momentum is; two balls with equal momenta (or opposite, or any other particular relationship) will have no special influence on each other.

This fact about interactions in the real world is reflected in the Hamiltonians that end up describing real-world systems. In principle, the Hamiltonian of a physical system could be any function of x and p. But in practice, we find systems that are described by Hamiltonians that look like (4.1)—a standard kinetic-energy term proportional to

Billiard balls will interact when they come into contact with another ball or a bumper at the same location in space. Nothing special happens when two balls have the same momentum.

p^2, and a potential-energy term that depends only on x, but perhaps with a complicated functional form.

This pattern continues when we move to systems with multiple moving parts. Consider two objects described by positions x_1 and x_2 and momenta p_1 and p_2. Almost all the time, the Hamiltonian is going to end up looking like

$$H\left(x_1, p_1, x_2, p_2\right) = \frac{p_1^2}{2m_1} + \frac{p_2^2}{2m_2} + V\left(x_1, x_2\right). \qquad (4.9)$$

We recognize the first two terms on the right-hand side as the individual kinetic energies of the two objects. The potential energy $V(x_1, x_2)$ depends on the positions of both objects in some way. If we're talking about billiard balls, the potential will be zero unless the balls are literally hitting each other. If we're talking about two planets interacting via gravity, their potential energy will be zero when the distance between them is very large, and it becomes appreciable when they get closer. The important thing is that distance in space is what determines the strength with which things interact.

These are all rules of thumb that apply to typical real-world systems, although there are some exceptions. The point is that, while we are free to imagine all kinds of crazy Hamiltonians, real physical systems are not so arbitrary. Objects have their own kinetic energy, and they interact with other objects via potential energy that depends on how close they approach.

This is why positions are different from momenta, and therefore why we think we live in space and not in momentum space. In real-world Hamiltonians, position is the variable in which interactions are local.

ACTION AT A DISTANCE

There is one subtlety here that we shouldn't just gloss over. For billiard balls, interactions are precisely local in space: The balls have no influence over each other right up to the point where they come into contact. But for gravity, the interaction does seem to stretch over some distance, even if it grows weaker as the distance goes up. Is that really "locality" as we've defined it?

Good question. Newton and his contemporaries spent a lot of time fretting about it. Many were convinced that the gravitational force should be mediated by some kind of substance in between the objects, rather than stretching effortlessly across space. They had implicitly accepted the idea of locality, and they worried that "action at a distance" shouldn't be part of a well-formulated theory. Newton himself agreed, calling the notion "so great an absurdity that I believe no man who has in philosophical matters a competent faculty of thinking can ever fall into it." But it seemed to work, and Newton was nothing if not practical. So figuring out what ultimately was going on was something that he "left to the consideration of [his] readers."*

You'll be pleased to know that subsequent generations have

---

* Letter from Isaac Newton to Richard Bentley, February 25, 1692. The Newton Project, http://www.newtonproject.ox.ac.uk/view/texts/normalized/THE M00258/.

considered it, and for the most part we have it figured out. The idea is to realize that space isn't empty, but that instead it is filled with **fields**, one of which is responsible for gravity.

A field is basically a function of space itself: At every point, the field takes on a certain value. Depending on the kind of field we're talking about, that value might be a number, or a vector, or something more complicated. There is the electric field, the magnetic field, the gravitational field, and various others. In modern physics, fields are the fundamental building blocks of reality (to the best of our current knowledge).

From the field point of view, the sun doesn't magically reach across space to exert a gravitational pull on each of the planets. Rather, it influences the gravitational field at its location, and the value of the field at each location influences the value at all locations immediately nearby. By concatenating all of these influences together—calculus!— we can determine the strength of the gravitational pull.

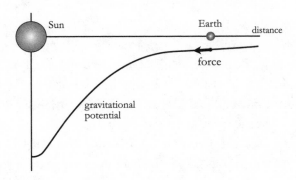

Fields and particles are complementary notions. A particle is characterized by its location in space: It is at this one place, and nowhere else. A field, by contrast, is everywhere; any field has a value at each spatial location. At any point in space, each field is being pushed around by other fields, but only by the values (and derivatives) of other fields *at precisely the same point in space*. The value of one field at point *A* has no direct influence on what happens at point *B*; there can only be an indirect influence, mediated by changes in the field that pass between the two points.

It was our old friend Pierre-Simon Laplace who first figured out how this could work for gravity. He posited the idea of a **gravitational potential field** filling space. This potential is related to the Newtonian gravitational force in the same way that the potential energy on a hilly landscape is related to the force on a rolling ball: The force is the derivative of the potential.

Laplace suggested two equations: one that determines how the values of the field respond to massive objects, and another that shows how the field pushed the objects around. In Laplace's picture, the sun creates a depression in the gravitational potential, which gradually relaxes as we move farther away from all of that mass. That relaxation implies a slope to the field, which the planets experience as the force of gravity. The predictions made by Laplace's theory are completely equivalent to those in Newton's, but with a different vocabulary to describe it.

Laplace's equations are precisely local: What happens at one point in space is only affected by what's happening immediately nearby. But one feature that is distasteful to modern sensibilities is that any change in the gravitational field will propagate instantly throughout the entire universe. If you pick up a bowling ball and move it from one place to another, the gravitational field caused by the ball slightly changes; if Newton and Laplace had been right, that change could be detected right away by a sufficiently sensitive detector on the other side of the galaxy.

These days, having been trained by Einstein's theory of relativity, that kind of instant transmission of information seems wrong to us. Signals shouldn't be able to travel faster than the speed of light. And indeed, Einstein was eventually able to replace the Newton/Laplace picture of gravity with his own theory, general relativity. General relativity once again posits a field in spacetime that explains the force of gravity, but it's a bit more mathematically complicated. Happily, it's as local as we could ever wish: The field itself is influenced only by what is happening in the neighborhood of each point, and any change in the field spreads only at the speed of light or slower.

One last point about locality before we move on. In modern physics, the phrase "action at a distance" is often modified by the adjective "spooky," in the memorable phrasing of Albert Einstein. By the time the twentieth century had rolled around, locality was pretty well ingrained in how physicists think about the world, so any sort of action at a distance would have been considered spooky. General relativity managed to rid our understanding of gravity of any such worries. But general relativity is still a classical theory, as opposed to a quantum one. What Einstein realized was that the advent of quantum mechanics opened the door to a new kind of action at a distance, resulting from measurements of entangled quantum particles. The effect he proposed is real and has been verified experimentally. But nature is sufficiently subtle that no useful information ever travels faster than the speed of light. So quantum mechanics leaves us in a weird situation where we have to use nonlocal concepts to explain what we observe but can't take advantage of them to send signals across space.

FIVE

TIME

We motivated our consideration of "space" by noting that it is the arena in which things happen. But "happen" implies transformation over **time**—something hadn't yet happened, then it was happening, then it had happened. If there were no time, there would be no motion, no evolution, no change. Everything interesting about the universe owes its existence to the passage of time.

Sometimes we speak as if time is some ineffably mysterious thing. In Saint Augustine's famous formulation, "If no one asks me, I know what time is. If I wish to explain it to he who asks, I do not know." But there's no mystery involved in how we use the concept of time in ordinary practice. If someone tells you, "Let's meet at eight o'clock," or "this podcast episode is ninety minutes long," we know exactly what they mean. The universe—space, and all the stuff within it—happens, over and over again, with slight changes, in a sequence of moments. Time is a label on those moments and a way of measuring the duration of some set of moments.

But if time itself is less mysterious than we might sometimes pretend, it certainly has mysterious aspects. How are time and space similar, and how are they different? Why is the past different from the future? Is the future fixed, or not yet determined? These are all excellent questions, to which we can provide answers at different levels of confidence.

TIME AND SPACE

Time is tricky because in some ways it's like space, and in other ways it's not. Let's ground ourselves by first considering the ways in which time is like space.

Time is part of how we locate ourselves in the universe. If you want to meet someone for coffee, you must specify both *where* and *when*. We could encode that information in four numbers: three to pick out a location in three-dimensional space, and one to specify a single moment of time. In practice we often leave some of that information implicit—we'll say "the coffeeshop downtown," assuming our friend knows both the address and the height above ground—but in principle it's there. If we told someone where to meet but not when, or vice versa, it wouldn't be of much use.

Once we think of time along with space as ways of locating ourselves in the universe, it's natural to group them together in a bundle we call "spacetime." You may have heard that the concept of spacetime was invented in the early twentieth century as part of the theory of relativity. That's true as a matter of historical fact. But even in Newtonian physics, before relativity came along, you *could* have lumped space and time together as "spacetime." The difference is just that the division of spacetime into "space" and "time" is completely fixed within Newtonian physics—there's a unique, right way to do it— whereas in relativity different observers will generally divide spacetime into space and time in different ways. In Newtonian physics, time and space are each absolute in their own right, while in relativity they are, you know, relative. (In particular, they are relative to our choice of a reference frame in spacetime.)

Even though Newtonian physics doesn't force us to lump space and time together, we're still welcome to do so. Once we do, it makes sense to draw a map—a simplified representation of the universe that includes both spatial dimensions and the single dimension of time. Such **spacetime diagrams** are a crucial part of the conceptual tool kit of the modern physicist. Generally, they are drawn with space stretching horizontally and time vertically, with the past toward the bottom and time increasing as we move upward.*

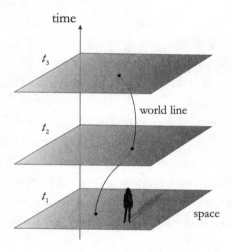

In a spacetime diagram, an object will not be represented by a single dot signifying its position but by a **world line** stretching from the past to the future, signifying its various positions at different moments in time. Your body takes up a three-dimensional volume in space at any one moment, but your life's history describes a four-dimensional

---

* At least, among people who specialize in relativity and spacetime. Computer scientists typically draw time increasing downward, while particle physicists have it go from left to right. Or perhaps from right to left, if they speak a language that is written in that direction.

worm extending through spacetime. Every physicist has, at one time or another, visualized their own world line from the moment of their birth to the current time.

LENGTH AND DURATION

The other obvious similarity between time and space is that you can measure them. Paths through space have a measurable length, and intervals of time have a measurable duration.

In practice, we can measure things when we have reliable standards against which to compare them. In the case of length, we have rulers. Take two solid rods of the same length, then let some time pass while protecting them against damage. Maybe one rod stays locked in a cabinet and another one goes on a trip for a while. When the second rod returns, we compare them again and find that they are still the same length. It's the reliability and reproducibility of this comparison that gives us a useful standard of length. We can pick one rod, call it the "standard meter," and then measure all other lengths with respect to it. Ultimately the existence of such marvelous artifacts can be traced to the underlying laws of physics. Rods are made of atoms, and atoms have basically fixed sizes, set by physical constants such as the masses and charges of electrons and atomic nuclei. Don't let anyone accidentally (or maliciously) scrape off any atoms from your standard meter, and you're set.

Duration is the same story, except we measure it with clocks rather than rods. A clock can be thought of as something that changes over time in a way that is reliable and reproducible in comparison to other clocks. This might seem circular, since we're using the notion of a clock to define itself. But that's no different from measuring rods: They are useful to the extent that they stay the same length relative to each other.

The point is that the real universe is full of clocks: systems that carry out some motion in predictable, regular ways with respect to one another. That's a feature of the laws of physics that allows us to measure the passage of time. (It's fun to try to imagine what things would be like if there were no clocklike objects in the universe.)

A classic example is the rotation of the Earth and its revolution around the sun. We can be fairly confident that over the course of one year, the Earth will rotate around its axis a bit more than 366 times.* It's this reliability that makes for good clocks.

Another example is the simple harmonic oscillator. If an oscillator is truly harmonic—the restoring force is precisely proportional to its displacement from equilibrium—it will take exactly the same amount of time to undergo a full oscillation, no matter what the amplitude of that oscillation is. The oscillator moves faster when it's oscillating by a lot, and slower when it's just a little, so that the period is the same either way. This makes harmonic oscillators an excellent starting point for building clocks.

Exact harmonic oscillators are hard to come by in nature. There is often some small effect that nudges a real physical system away from being exactly harmonic. But fortunately, we've already seen that oscillating systems tend to be *approximately* harmonic when their oscillations are sufficiently small. There is a famous story about Galileo, related by his first biographer, according to which he noticed that a lamp hanging in the cathedral in Pisa would swing with a fixed period. This gave him the idea that a pendulum could serve as the basis for an accurate clock. Later in life (in fact, after he had gone blind), Galileo returned to this idea, and with the help of his son he proposed how such a clock might work. They didn't get to complete the project,

---

* Why not 365? Because even if the Earth didn't rotate at all, the sun would still rise and set once per year, because of the Earth's revolution around it. And its motion would be backward from what we see because of rotation. We end up seeing one fewer sunset per year than the number of complete rotations of the Earth.

but soon thereafter Christiaan Huygens developed the first practical design.

A pendulum, however, is not a perfect harmonic oscillator; when its swings are too large, they don't have quite the same period as when they are small. To ensure accuracy, you really want to base your clock on a pendulum that swings only a little bit, so you can take advantage of the fact that all oscillators behave harmonically at small amplitudes. Robert Hooke ultimately invented a gizmo called the anchor escapement, which allowed for the construction of practical pendulum clocks that oscillated just a few degrees at a time.

EVOLUTION THROUGH TIME

That's the easy part; the ways that time is like space. But it's not exactly like space. At a visceral level, time seems to *happen* in a way that doesn't quite apply to space. We feel like time flows, or that it passes us by. We're not tempted to describe space as flowing around us, or us through it. To our untutored intuition, at least, space seems like a collection of things (points, one at each location), whereas time seems like some kind of process.

Like many intuitive beliefs, this feeling that time and space are quite different things doesn't come from nowhere. In time, for example, conditions at one moment evolve smoothly into those at the next. Think of Newton's second law, equating force to mass times acceleration (which in turn is the second derivative of position):

$$\vec{F} = m\vec{a} = m\frac{d^2\vec{x}}{dt^2}. \qquad (5.1)$$

Or, just as well, consider Hamilton's equations for the time derivative of position and momentum:

$$\frac{dp}{dt} = -\frac{\partial H}{\partial x}, \quad \frac{dx}{dt} = \frac{\partial H}{\partial p}. \qquad (5.2)$$

In either case, these relations are meant to hold at each moment in time. They determine what happens at other times.

Nothing like that works for space. If you find a heavy rock—too heavy to move with any feasible means at your disposal—chances are very good that the rock will still be there a moment later in time. But if you find a heavy rock at some particular location, there's no rule saying there needs to be another rock (or more of the same rock) at a nearby location in space. As we consider different positions in space, conditions can change quite dramatically from point to point. But as we consider moments in time, one condition flows directly into another.

THE ARROW OF TIME

The apparent "flow" of time seems—again, at an intuitive level—to have a pronounced directionality to it. Time seems to move from the past to the future, not the other way around. This feature is the famous **arrow of time**, so named by British astrophysicist Arthur Eddington in 1927.

The arrow of time is fundamentally an asymmetry: From the point of view of "now," the past and future have quite different properties. The past is fixed, it's in the books, while the future seems open, up for grabs. We have memories and records of the past: photographs, fossils, history books, artifacts in the present that give us some definite knowledge of what happened at some earlier time. There is nothing like that for the future. We can try to predict it, but we don't have any pictures of it. We like to think that our choices here and now can influence what the future will become, but most of us admit that our choices today don't influence yesterday.

The arrow is such an indelible part of how we think time works that early thinkers wouldn't even have thought of it as something that required explanation. *Of course* the past and future are different; how could they be otherwise? They are fundamentally different things.

The advent of classical mechanics turned the arrow of time into a bit of a mystery. It's there that we learn that information is conserved. From the present state of an isolated system, we can predict its future and equally well retrodict its past. This feature is known as **reversibility**: If the laws of physics imply that a system in state A at time 1 evolves to be in state B at time 2, those same laws imply that if the system were in state B at time 2, it must have been in state A at time 1. We can run the clock either forward or backward equally well.

Reversibility isn't manifest in our everyday experience. A glass of warm water with ice in it will, given a bit of time, evolve to a glass of room-temperature water. But a glass of initially room-temperature water will also evolve to a glass of room-temperature water (that is, it won't really change at all). Therefore, if you see a glass of water at room temperature, you don't know whether it used to be warm water with ice or whether it's been at room temperature all along.

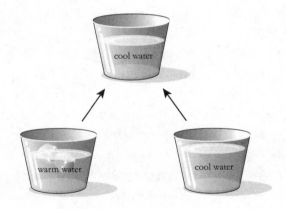

So classical mechanics implies secret reversibility: The underlying laws of physics are reversible even if our experience of the world around us is certainly not. We sometimes say that the "microscopic" or "fundamental" dynamics are reversible, but the "macroscopic" or "emergent" dynamics aren't. Here "microscopic" usually means "a small number of moving parts," rather than "small in size." Planets in the solar system are very well described by reversible Newtonian mechanics, even though they're pretty big. The question becomes why the macroscopic world seems irreversible at all.

TIME REVERSAL

Before digging into the origin of irreversibility, we should examine the similar-sounding (but ultimately different) notion of **time-reversal** symmetry. It's potentially confusing, but worth getting straight since you will hear people talking about it and pretending it has some connection to the arrow of time (it doesn't, not really).

A **symmetry** is a transformation that switches around some of the variables of the system but leaves the essential story intact. The time-reversal transformation T involves taking the time variable t and multiplying it by -1 to reverse its direction:

$$t \rightarrow -t. \tag{5.3}$$

Under this transformation, the laws of classical physics are **invariant**—they remain unchanged. This feature of time-reversal invariance is closely related to reversibility, but it's not quite the same thing.

Think about Newton's second law, as written in (5.1). The mass of the object doesn't change under time reversal. Nor does the force, at least for well-known examples like the force due to gravity or the slope of a hill. What about the acceleration, $\vec{a}$? The position $\vec{x}$ doesn't change. The infinitesimal time interval dt does; since $t \rightarrow -t$, we also have $dt \rightarrow -dt$. (If you reverse the direction of time, you reverse the direction in which time increments.) But the acceleration involves the square of that, $\vec{a} = d\vec{v}/dt = d^2\vec{x}/dt^2$, and we have $(dt)^2 \rightarrow (-dt)^2 = (dt)^2$. So the acceleration remains unaltered by the transformation (5.3). Hence, the entirety of Newton's second law is invariant under time reversal.

The same thing holds for the Hamiltonian version of classical mechanics, but in a slightly more fun way. There are two tricks you need to keep in mind. First, when you do time reversal you should send momentum to minus itself, $\vec{p} \rightarrow -\vec{p}$. You might think this goes without saying, since momentum is mass times velocity, and velocities reverse when we run time backward. But momentum isn't defined by mass times velocity in the Hamiltonian picture; it's an independent variable that happens to equal mass times velocity on trajectories that solve the equations of motion. Nevertheless, you should reverse its sign if you want time-reversal symmetry to be obeyed. (This is a clue that something subtle is going on with this whole business.)

The second trick to keep in mind is that the Hamiltonian is kinetic energy plus potential energy, and kinetic energy usually looks like $\vec{p}^2/2m$. Because the momentum is squared, the kinetic energy itself remains invariant when we reverse the sign of the momentum.

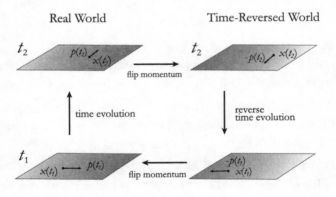

The fundamental equations of classical mechanics, in other words, don't know the difference between moving toward the past and moving toward the future.

REVERSIBILITY AND CPT

"Reversibility" and "time-reversal invariance" are two ideas that sound alike, but there's a subtle difference. Reversibility says that information is conserved over time, so that systems don't forget what state they started in; time-reversal invariance says that the laws of physics look the same running forward or backward in time. Reversibility, it turns out, is more fundamental: As long as the dynamics are reversible, we can always cook up some notion of time-reversal symmetry.

We can approach this issue by way of a detour. If you hang out with the wrong crowds—in this case, with particle physicists—you might be told that in the world of elementary particles, time reversal is not a symmetry of fundamental physics after all. We've even done experiments that demonstrate the violation of time-reversal invariance. Under precisely the right circumstances you can take a collection of particles in some state A, evolve them through time to some different state B, then reverse all the momenta to run them "backward in time," and they do *not* end up precisely in state A as we might have expected.

That's not the only symmetry you might have expected to find in nature, but which is actually violated. In addition to time reversal, we also have space reversal, more often known as **parity**, and denoted P. A parity transformation reverses all three dimensions of space at once; it's equivalent to looking at a system in a mirror, so that writing appears backward and right-handed screws look like left-handed screws. Elementary particles often have spin, and a parity transformation would reverse the direction in which they are spinning. Parity was originally believed to be a good symmetry of nature, but in experiments in the 1950s C. S. Wu and others showed that it is violated.

Then there is a symmetry called **charge conjugation**, or C. Many species of elementary particles have associated kinds of **antiparticles**. The antiparticle of an electron gets its own name, the positron, but for the most part we just stick on the prefix anti-, so that the antiparticle of a neutrino is an antineutrino. Charge conjugation exchanges particles for their antiparticles. Like T and P, it turns out to be violated by certain particle interactions. Charge conjugation might seem like a different beast from time or space reversal, but the direction of time is intimately connected with the distinction between matter and antimatter—there is a sense in which antiparticles are mathematically equivalent to particles traveling backward in time.

Here's the punch line: Individually, C, P, and T are all violated in elementary-particle physics. But the combination of transformations, CPT, is conserved. There are good theoretical reasons for this to be true, and thus far it's been experimentally confirmed. If we take an arrangement of particles in state A and evolve them forward in time to state B, we can construct the mirror image of B, switch particles with antiparticles, reverse all the momenta, then evolve backward in time, and once again take the mirror image. The result will be our original state A. That's CPT symmetry in action.

This suggests the following bold strategy: Maybe we could just *define* "time reversal" to be what we've just called CPT? In other words, could we define a new operation, call it T', by first doing a naïve time reversal (T), then a parity transformation (P), then a charge conjugation (C), and think of that new operation $T' = CPT$ as "what we really mean by time reversal"? This new souped-up version of time reversal would then be a symmetry of nature after all.

Sure, we can do that. In fact, it turns out that we can always define a time-reversal operator by changing the direction of time plus appropriate alterations to other variables, so that the result is a good symmetry—*if* the underlying theory is reversible. Reversibility implies some notion of time-reversal symmetry, even if it might not look precisely like the naïve one we originally guessed.

Recall that even in conventional Hamiltonian mechanics we had to reverse the direction of momentum to implement time reversal, but we didn't really offer a principled justification for taking that step. Now we know why: Defining time reversal so that it's a good symmetry of the dynamics will generally involve some extra manipulations, in addition to just sending $t \rightarrow -t$. Sometimes those extra steps seem natural and even inevitable, like sending $\vec{p} \rightarrow -\vec{p}$, so we just bundle them up with the definition of "time reversal" and think no more about it. Other times they might seem ad hoc, like throwing in parity and charge conjugation, so we mumble something about "time reversal being violated, but this closely related symmetry is still preserved." The physical point, independent of our definitional choices, is that some such symmetry is there whenever dynamics are reversible.

So the breakdown of naïve time-reversal invariance T has nothing to do with the arrow of time. That feature of particle physics, while important in its own right, leaves reversibility intact. The arrow of time stems from the fact that the macroscopic world does not appear reversible, even though the microscopic world seemingly is.

ENTROPY

What, then, is responsible for irreversibility, and thus for time's arrow? The ultimate answer lies in the fact that the **entropy** of a closed system, including the universe as a whole, tends to increase over time. Entropy is often roughly defined as the disorderliness or disorganization of a system—a deck of cards in perfect order has low entropy, while a randomly shuffled deck has high entropy. That's good enough for most purposes, but we can be a little more precise.

To Laplace's Demon, who keeps track of the exact microscopic state of the world, everything would seem reversible. Then again, the Demon also knows the complete past and future. The Demon draws no distinction between "remembering the past" and "predicting the future." For the Demon, there's no real arrow of time.

But none of us is Laplace's Demon. We're nowhere close to being Laplace's Demon. We are finite human creatures, with dramatically limited capacities of observation and calculation. We have trouble remembering a phone number, much less the positions and momenta of 6×10^{23} or more particles.* We don't know everything about the exact state of the world, and we don't even see the exact state of the world when we look at it. When we glance around a room, we see tables and chairs and people; we don't precisely measure the position and momentum of each elementary particle of which they are made.

What we do, instead, is what we call **coarse-graining**: We group a bunch of specific states of a system into a single description, then use that description to understand what's going on and make the best predictions we can. When we describe a box of gas or a cup of coffee, we can specify the temperature and pressure and velocity of the fluid at each point in the container—the **macrostate** of the system—but there will still be many specific arrangements of atoms and molecules—**microstates**—fitting that description. But we don't need to know the exact microstate to know that gas will expand to fill a box, or that hot coffee will cool down to room temperature over time. We can do pretty well making predictions on the basis of just knowing what macrostate we have.

Giving a precise definition to macrostates is a subtle business, but the basic idea is "the set of all microstates that look the same to a macroscopic observer." Some macrostates, like when a collection of gas molecules is distributed uniformly through a box, correspond to a large number of possible microstates. Meanwhile, other macrostates, like when the gas molecules just happen to all congregate in one corner of the box, correspond to a relatively small number of possible microstates.

---

* The modern laptop computer on which I am writing this has 64 GB of RAM, or enough to store very approximate positions and momenta of about 10^9 particles. You would need approximately a million billion such computers to store the information in a relatively small macroscopic system, even if you devoted all of their memory to the task. It's not going to happen anytime soon.

It was the brilliant idea of Austrian physicist Ludwig Boltzmann in the 1870s to use this setup to suggest a way of understanding entropy. The idea of entropy had already been introduced earlier in the nineteenth century; Boltzmann's contribution was to connect this intrinsically macroscopic property to its microscopic underpinnings. Namely, he suggested that the entropy is related to the number of microstates in each macrostate.* From this perspective, it makes sense that entropy tends to increase over time. A low-entropy macrostate corresponds to just a small number of possible microstates, whereas a high-entropy macrostate corresponds to a large number of possible microstates. If we start in a low-entropy state and our system wanders off in some generic direction in phase space, we should expect entropy to increase simply because there are more ways (usually many, many more ways) to be high-entropy than to be low-entropy.

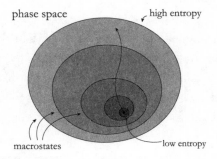

This is the **second law of thermodynamics**—in closed systems, entropy will either increase or stay constant, never spontaneously decrease. (The first law is just energy conservation.) Many systems are not closed, in that they interact with the external world. In that case, entropy can certainly decrease; when you put a bottle of Champagne in the refrigerator, it cools off and its entropy goes down. But you'll

---

* Specifically, the entropy of a macrostate is proportional to the logarithm of the number of microstates therein. See Appendix A for a discussion of logarithms.

notice that the refrigerator is expelling heat from the back. Entropy as a whole is still increasing. That's why you can't cool off a room by plugging in a refrigerator and leaving its door open. The laws of thermodynamics won't let you.

Boltzmann's insight provides a connection between the micro-world of particles and Laplace's Demon, and the macroworld of coarse-graining, approximation, and incomplete information. The fact that we can say anything useful at all about physical systems while remaining ignorant of so many details about their microscopic states is a feature of **emergence**. We'll put aside for now the important question of why the world admits higher-level emergent descriptions, but it's one to keep in the back of our minds.

THE PAST HYPOTHESIS

Entropy tends to increase because there are more ways (microstates) to be high-entropy than to be low-entropy. You may have noticed that this explanation smuggles in a hidden assumption: that the entropy started low in the first place. This idea, known simply as the **past hypothesis**, works together with Boltzmann's micro/macro connection to get the second law of thermodynamics off the ground.

We can find some clarity by once again comparing time with space. While there is clearly an arrow of time, there is no such thing as the "arrow of space." You can tell the difference between "up" and "down," but that's only because we live in the vicinity of an influential object—the Earth—and feel its gravitational pull. If you were an astronaut floating in a space suit, there would be no noticeable difference between up and down, left and right, forward and back.

But wait a minute. We *are* here on Earth, and there *is* an arrow that distinguishes up from down. It's just not built into the laws of physics; it's an accident of our local environment. It's due to the particular configuration of matter in our vicinity—a big, influential planet. Could the same thing be true for time?

It could be, and in fact it is. Because the underlying laws of physics are reversible, if all we knew was that the current entropy of the universe were low, we would predict that it would be higher in the future—but we'd also predict that it had been higher in the past. We would be living at some special historical moment when entropy just happens to be anomalously low.

Nobody believes that's true. The fundamental symmetry between past and future is broken by a boundary condition on the history of our universe, or at least our observable part of it. That boundary condition sets the initial entropy to be very low indeed.

Our universe is expanding, and about 14 billion years ago, everything we currently see was squeezed into a very hot, very dense, rapidly expanding state. If we extrapolate all the way back, our best current theory (Einstein's general relativity) predicts a "singularity" of infinite density, labeled "the Big Bang." The right way to think about that is not that there really was such a singularity but that the theory itself breaks down. Someday we'll have a better understanding of gravity and the expanding universe, which will hopefully reveal what happened at the moment we call the Big Bang.

For right now, we can content ourselves with describing what happened just after that moment, about which both theory and observation have told us a lot. The universe consisted of a hot, dense plasma, distributed uniformly throughout space.

That might sound like a high-entropy configuration; in a box of gas in a lab here on Earth, the highest-entropy configuration is one where the gas is distributed uniformly. But there is a crucial difference when we're talking about the universe as a whole, which is so big and massive that gravity becomes crucial. Gravity works to disturb that smoothness, pulling matter into dense regions and emptying out less-dense ones. That's precisely what happens over the next few billion years as stars and galaxies and clusters coalesce out of the initial smoothness. And during that whole process, entropy is increasing.

When gravity is important, there are more ways for particles to be in a lumpy distribution than in a smooth one. So the initial state of the universe had an enormously smaller entropy than we would expect if it were (somehow) randomly chosen from all possible configurations.

We don't know why the universe started in a low-entropy configuration. After all, low-entropy configurations are supposed to be special; they correspond to far fewer microstates than high-entropy ones. That's a question for cosmology that we'll put aside for now.

What matters is that our observed universe did start out in a low-entropy state, and entropy has been going up ever since. That's the ultimate origin of the direction of time. Just as there is an arrow of space nearby because we live close to the Earth, there is an arrow of time because we live close (relatively speaking) to the Big Bang.

To be fair, much remains to be explained. The arrow of time manifests itself in many ways: memory, causality, aging, and more. We've been talking about the "thermodynamic" arrow of time, with the idea that it ultimately underlies all these other directional aspects—but we haven't carefully demonstrated that. These issues are still research-level open questions.

PRESENTISM, ETERNALISM, POSSIBILISM

We've painted a picture in which the arrow of time isn't built into the fundamental laws of physics but is an epiphenomenon—a by-product of the fact that the universe started in a condition of low entropy, and that low-entropy states tend to evolve into increasingly higher-entropy ones. The directionality or flow of time isn't part of the fundamental architecture of reality, which makes no distinction between evolution to the past or future, nor between different moments of time.

This would be a good place to note that not everyone agrees. The story we've told is probably the closest thing we have to a "standard" view among scientists and philosophers who have thought about these things, but it's not universally accepted. The physics here is pretty well established, but there are alternative philosophical positions to consider (that might someday feed back into an improved understanding of the physics, if they turn out to be right).

At issue is, once again, what we think of as "real." In our discussion of space, we distinguished between the idea that space is a separate substance in its own right and whether it's simply a way of organizing relations between objects. But we weren't tempted to think that some locations in space are real while others are not. When it comes to time, that temptation is front and center. Traditionally, in fact, it's been most common to only think of the current moment of time, the "now," as truly real. The past is gone, according to this view, while the future has yet to occur. So they're not real in the same way that the present is.

The view that only the present moment is real is called **presentism**, sensibly enough. It can be contrasted with **eternalism**, the view that all moments of time are equally real, which we implicitly assumed in some of the discussion above. Eternalism is also called the **block universe** view, since it takes the real world to be the whole four-dimensional block of spacetime.

Presentism and eternalism aren't the only options on the table. Those who are tempted by presentism may also be impressed by the difference in how we think about past and future. You can't change the past, after all, whereas we think our actions here and now can have an impact on the as-yet-unformed future. So a third tempting possibility is **possibilism**, or the "growing present" view. There, we count both the past and the present as real while denying that status to the future.

We're not going to offer any final adjudication in the dispute between these perspectives, but we can acknowledge that there are sensible motivations behind each of them. Presentism comes closest to our intuitive, pre-scientific view of the world. Eternalists look at the laws of physics, notice that they are reversible and conserve

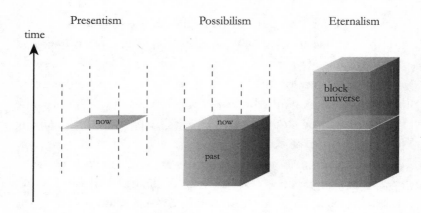

information, and also that they do not distinguish any special moment as "now." Thus, they say, we should attach equal status to past,
present, and future. We can explain our impressions that time flows
and that there is something more real about now than about other
moments as consequences of the arrow of time, which is itself a consequence of increasing entropy.

To this, presentists will shake their heads, lamenting that their
eternalist friends must twist themselves into knots to explain what is
right in front of their nose. Wouldn't it be easier, they say, to explain
the flow of time by positing that it actually does flow? That only the
now is real, and time is how we talk about the process of change, not
merely a coordinate on some entirely intangible four-dimensional reality? Indeed, say the possibilists, and as long as we're taking our intuitions seriously, let's not be so blinkered as to put the past and future
on an equal footing.

BEING AND BECOMING

These are modern-day descendants of ancient disputes. Presentism is
often traced to the Greek philosopher Heraclitus, who emphasized
the primacy of change: You can never walk into the same river twice,
since at a different time it will be a different river. A more eternalist

view can be found a generation later in Parmenides, who conceived of an eternal universe without end. Parmenides, in other words, prioritized **being**, while Heraclitus put the concept of **becoming** at the top of the metaphysical heap.

The being/becoming distinction is reflected not just in the eternalism/presentism dichotomy but in how we think about the laws of nature themselves. According to one school of thought, laws are just convenient ways of summarizing what happens in the world. We could imagine specifying the universe by listing what happens at every single moment of time, but it's simpler (though still incalculably hard) to specify what happens at just one moment, then say "and other times are related to that one by the laws of nature." This view is dubbed **Humeanism**, after Scottish philosopher David Hume, though it's not clear that Hume himself would have qualified as a Humean in this sense.

The alternative is known as **anti-Humeanism**, presumably because no famous philosopher has championed the view loudly enough to get their name associated with it, although it does have plenty of adherents.* Anti-Humeans think that the laws of nature don't merely describe the world; they act to bring the world into existence, one moment after another. The laws, in other words, don't just go along for the ride; they have an independent existence and are responsible for the universe being the way it is.

Humeanism fits comfortably with an eternalist sensibility, while anti-Humeans are more likely to be presentists. To see how these viewpoints are related, consider **counterfactual** statements—claims about what the world would have been like had it been a bit different.

---

* According to a recent survey, 31 percent of philosophers take a Humean view of the laws of nature, 54 percent are anti-Humean, and 15 percent reply "other." D. Bourget and D. Chalmers (2021), "Philosophers on Philosophy: The 2020 PhilPapers Survey," https://philarchive.org/archive/BOUPOP-3/.

("I would have skipped the appetizer had I known the dessert selection was going to be this irresistible.") Counterfactuals can be thought of as statements about other **possible worlds**—sensible versions of reality that just happen to not be the reality in which we find ourselves. We deploy them all the time; without thinking about other possible worlds, we couldn't talk about choice, responsibility, morality, or other ideas to which we are generally quite attached.

But wait a minute. To the Humean, all that exists is the actual universe. What we call the "laws of nature" are just helpful summaries of patterns we see within *this* universe. So how can a Humean ever talk about counterfactuals? Or at least, how could they consistently imagine a *different* universe, but with the *same* laws of physics? Humeans think that laws simply describe the world, they don't govern it. Anti-Humeans don't have this worry. To them, laws don't merely summarize what happens, they govern what happens. It's easy to mentally extend that power to possible worlds with other initial conditions but the same laws.

However, anti-Humeans face the challenge of explaining what it *means* to say that the laws of physics "govern" what happens from one moment to the next. If we knew everything that ever happened in the universe, it would be hard to pinpoint exactly what additional thing we were missing. Humeans often have a sneaking suspicion that anti-Humeans are still operating under an antiquated crypto-Aristotelian view, where parts of the world have essences and natures. Maybe it's okay to think that things just happen, and to take as our task describing what those things are.

SIX

SPACETIME

In Newtonian physics, you would have been welcome to consider space and time together as a single four-dimensional **spacetime** in which events are located. But nothing would have forced you to do that; space and time had their independent identities, and nobody ever got them mixed up. It was with the theory of relativity, put together in the early twentieth century, that talking about spacetime became almost unavoidable. In relativity, it's no longer true that space and time have separate, objective meanings. What really exists is spacetime, and our slicing it up into space and time is merely a useful human convention.

One of the major reasons why relativity has a reputation for being difficult to understand is that all of our intuitions train us to think of space and time as separate things. We experience objects as having extent in "space," and that seems like a pretty objective fact. Ultimately it suffices for us because we generally travel through space at velocities far less than the speed of light, so pre-relativistic physics works pretty well.

But this mismatch between intuition and theory does make the leap to a spacetime perspective somewhat intimidating. To make things worse, presentations of relativity often take a bottom-up approach, starting with our everyday conceptions of space and time and seeing how they become altered in the new context of relativity. We are told stories of stretching rods and distorted clocks, all of which are technically accurate but can obscure the beauty and simplicity of the underlying concepts.

We're going to be a little different. Our route into special relativity might be thought of as top-down, taking the idea of a unified spacetime seriously from the get-go and seeing what that implies, rather than starting from more familiar-seeming notions and seeing how they become modified in this new context. We'll have to stretch our brains a bit, but the result will be a much deeper understanding of the relativistic perspective on our universe.

THE BIRTH OF SPACETIME

The development of relativity is usually attributed to Albert Einstein, but he provided the capstone for a theoretical edifice that had been building since James Clerk Maxwell unified electricity and magnetism into a single theory of **electromagnetism** in the 1860s. Maxwell's theory explained what light is—an oscillating wave in the electromagnetic fields—and seemed to give a special place to the speed at which light travels. The idea of a field existing all by itself wasn't completely intuitive to scientists at the time, and it was natural to investigate the question of what was actually "waving" in a light wave.

Various physicists investigated the possibility that light propagated through a medium they dubbed the "luminiferous aether." But nobody could find evidence for any such aether, so they were forced to invent increasingly complicated reasons why this substance should be undetectable. Einstein's contribution in 1905 was to point out that the aether had become completely unnecessary, and we could better understand the laws of physics without it. All we had to do was to accept a completely new conception of space and time. (Okay, that's a lot, but it turned out to be totally worth it.)

Einstein's theory came to be known as the special theory of relativity, or simply **special relativity**. His foundational paper was titled "On the Electrodynamics of Moving Bodies," in which he argued for new ways of thinking about length and duration. He explained the special role of the speed of light by positing that there is an absolute speed limit in the universe—a speed at which light just happens to travel when moving through empty space—and that everyone would measure that speed to be the same, no matter how they were moving themselves. To make that work out, he had to alter our conventional notions of time and space accordingly.

But he didn't go quite so far as to advocate joining space and time into a single unified spacetime. That step was left to his former university professor, Hermann Minkowski, in 1907. The arena of special relativity is today known as **Minkowski spacetime**.

Once you have the idea of thinking of spacetime as a unified four-dimensional continuum, you can start asking questions about its shape. Is spacetime flat or curved, static or dynamic, finite or infinite? Minkowski spacetime is flat, static, and infinite. Einstein worked for a decade to understand how the force of gravity could be incorporated into his theory. His eventual breakthrough was to realize that spacetime could be dynamic and curved, and the effects of that curvature are what you and I experience as "gravity." The fruits of this inspiration are what we now call **general relativity**.

So special relativity is the theory of a fixed, flat spacetime, without gravity; general relativity is the theory of dynamical spacetime, in which curvature gives rise to gravity. Both count as "classical" theories even though they replace some of the principles of Newtonian mechanics. To physicists, "classical" doesn't mean "non-relativistic"; it means "non-quantum." All of the principles of classical physics, including our ability to think of things in terms of Hamiltonian dynamics or the principle of least action, are fully intact in the relativistic context, even if we have a somewhat more sophisticated notion of space and time.

TWO NOTIONS OF TIME

Special relativity is often introduced in terms of "length contraction" and "time dilation." These are perfectly respectable concepts, but they reflect an old-fashioned tendency to treat space and time as separately fundamental, albeit stretchier than we are used to.

The truth is more profound. We should be willing to let go of our pre-relativity fondness for the separateness of space and time, and allow them to dissolve into the unified arena of spacetime. The best way to get there is to think even more carefully about what we mean by "time." And the best way to do that is to harken back, once again, to how we think about space.

Consider two locations in space, such as your home and your favorite restaurant. What is the distance between them?

Well, that depends, you immediately think. There is the distance "as the crow flies," if we could imagine taking a perfectly straight-line path between the two points. But there is also the distance you would traverse on a real-world journey, where perhaps you are limited to taking public streets and sidewalks, avoiding buildings and other obstacles along the way. The distance you traverse is always going to be longer than the distance as the crow flies, since a straight line is the shortest distance between two points.

Now consider two **events** in spacetime. In the technical jargon of relativity theory, an "event" is just a single point in the universe, specified by locations in both space and time. One event, call it A, might be "at home at 6 p.m.," and event B might be "at the restaurant at 7 p.m." Keep these two events fixed in your mind, and think about a journey between A and B. You can't hurry to get to B sooner; if you arrive at the restaurant at 6:45, you will have to sit around and wait until 7 p.m. to reach the event in spacetime we have labeled *B*.

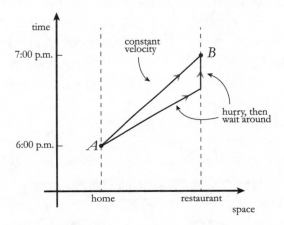

Now we can ask ourselves, just as we did for the spatial distance between home and restaurant, how much time elapses between these two events?

You might think this is a trick question. If one event is at 6 p.m. and the other is at 7 p.m., there is one hour between them, right?

Not so fast, says Einstein. In an antiquated, Newtonian conception of the world, sure. Time is absolute and universal, and if the time between two events is one hour, that's all there is to be said.

Relativity tells a different story. Now there are two distinct notions of what is meant by "time." One notion of time is as a **coordinate on spacetime**. Spacetime is a four-dimensional continuum, and if we want to specify locations within it, it's convenient to attach a number called "the time" to every point within it. That's generally what we have in mind when we think of "6 p.m." and "7 p.m." Those refer to values of a coordinate on spacetime, labels that help us locate events. Everyone is supposed to understand what we mean when we say "meet at the restaurant at 7 p.m."

But, says relativity, just as the distance as the crow flies is generally different from the distance you actually travel between two points in space, the duration of time that you experience along your world line generally won't be the same as the universal coordinate time. You experience an amount of time that could be measured by a clock that you carry with you on the journey. This is the **proper time** along the path. And the duration measured by a clock, just like the distance traveled as measured by the odometer on your car, will depend on the path you take.

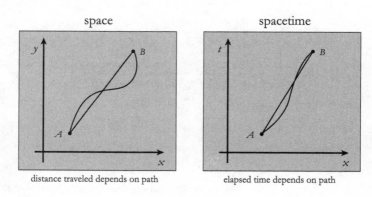

distance traveled depends on path   elapsed time depends on path

That's one aspect of what it means to say that "time is relative." We can think both about a common time in terms of a coordinate on spacetime and about a personal time that we individually experience along our path. And time is like space—those two notions need not coincide. (As historian Peter Galison has pointed out, it's not a coincidence that Einstein worked in a Swiss patent office at a time when rapid rail travel was forcing Europeans to think about what time it was at other cities across the continent, so that building better clocks became an important technological frontier.)

INERTIAL TRAJECTORIES ARE LONGEST TIMES

There must be some way in which time is not like space, otherwise we'd just talk about four-dimensional space, rather than singling out time as deserving of its own label. And we're not thinking of the arrow of time here—for the moment we're back to a simple world with few moving parts, where entropy and irreversibility aren't things we have to worry about.

The difference is this: In space, a straight line describes the shortest distance between two points. In spacetime, by contrast, a straight path yields the **longest elapsed time** between two events. It's that flip from shortest distance to longest time that distinguishes time from space.

By "straight path" in spacetime we mean both a straight line in space and also a constant velocity of travel. In other words an inertial trajectory, one with no acceleration. Fix two events in spacetime— two locations in space and corresponding moments in time. A traveler could make the journey between them in a straight line at constant velocity (whatever that velocity needs to be to arrive at the right time), or they could zip back and forth in a non-inertial path along the way. The back-and-forth route will always involve *more* spatial distance, but *less* proper time elapsed, than the straight version.

Why is it like that? Because physics says so. Or, if you prefer, because that's the way the universe is. Maybe we will eventually uncover some deeper reason why it had to be that way, but in our current state of knowledge it's one of the bedrock assumptions upon which we build physics, not a conclusion we derive from deeper principles. Straight lines in space are the shortest possible distance; straight paths in spacetime are the longest possible time.

It might seem counterintuitive that paths of greater distance take less proper time. That's okay. If it were intuitive, you wouldn't have needed to be Einstein to come up with the idea.

THE TWIN THOUGHT EXPERIMENT

There is a colorful illustration of this principle, usually known as the **twin paradox**, even though it is absolutely, indisputably not a paradox. It's just a nonintuitive feature of physics that we tend not to come across in our daily lives, largely because we tend to move at velocities much slower than the speed of light. It's bad form to attach words like "paradox" to phenomena that make perfect sense but seem unusual to us when we're first exposed to them.

Consider two twins, Alice and Bob. They don't even have to be twins—people who move on different paths through spacetime will experience different amounts of time regardless of whether they are related—but the scenario is more vivid when they are the same age. Alice stays put here on Earth, while Bob hops into an extremely advanced spaceship, zooms away at a velocity close to the speed of light, then turns around and zooms back to rejoin Alice. We will imagine that the acceleration of Bob's ship is essentially instantaneous, and he is unharmed by the enormous g-forces that would entail. We also ignore details like the fact that the Earth moves (slowly) through space, and that Alice is sitting in a gravitational field. Thought experiments let us invoke these kinds of spherical-cow approximations.

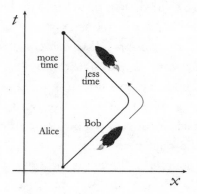

According to this scenario, Alice has moved on a straight line through spacetime by not moving in space at all. Bob's world line takes him from the same initial event to the same final event, but on a highly non-straight path, due to that turnaround at the midpoint.

(His path once he departed would have been straight if he hadn't turned around, but in that case he wouldn't have rejoined Alice at a later spacetime event, so there wouldn't be any way for them to compare their ages.)

What the twins find at their reunion is that Bob has experienced less proper time than Alice has. He's not as old as she is, and any clock he brought with him has counted off fewer minutes. Plugging in numbers, if Bob zoomed out and back at a constant velocity relative to Earth of 99 percent the speed of light, for every one year of time that he experienced, Alice will have experienced approximately seven years. (We'll empower you to do this calculation below.) If Bob spent a few years on his journey, he would return to find that his twin sister is considerably older than he is upon his return.

We haven't (yet) sent out any human astronauts on round-trip journeys near the speed of light to test this prediction. But we have done equivalent experiments with elementary particles, and the effect is indisputably real. The amount of time that travelers experience depends on the path they take through spacetime, exactly as the distance traversed through space depends on the path you walk through it.

None of which, as we've noted, is remotely paradoxical. But it is a radical shift in how we think about the nature of time. In Newton's universe, time was universal, and clocks wouldn't go out of sync just because someone took a rapid jaunt through space. In Einstein's universe, time is personal, and your clocks reflect the particularity of your journey.

THE SPEED OF TIME

The twin thought experiment has inspired a great deal of not-very-helpful discussion about the "rate at which time passes." If Bob doesn't age as much as Alice, it sure is tempting to say that Bob's clock was running more slowly. Giving into that temptation leads to all sorts of confusions.

Under certain circumstances—when we're excited, or bored, or quivering with anticipation, or experiencing sheer terror—it can seem like the rate at which time passes changes. Time can feel like it's speeding up or slowing down, depending on our situation. Rest assured, this feeling is entirely a matter of biology and psychology, not a matter of physics. Your internal clocks, unreliable biological contraptions that they are, have gone out of whack. Time itself cannot speed up or slow down.

After all, what would it mean for time to "speed up"? A speed is the rate at which something happens (the derivative!) *with respect to time*. The velocity at which you are traveling is the rate at which you cover distance as time passes. The rate at which a container fills up is the volume of liquid introduced over time. So the speed of time, if that concept were to make sense at all, would be the rate at which time passes with respect to time. To wit: one second per second. It literally cannot be anything else. It's not a quantity that makes any more sense to worry about than the number of meters you travel per meter traveled.

You will sometimes hear that time can speed up or slow down according to the theory of relativity. That's baloney. Or to be more polite about it, it's a misleading way to describe a real phenomenon.

The new feature in relativity is that the total duration experienced by two observers moving in different ways will generally not be equal, even if they begin and end at the same events in spacetime. This isn't because the rate of time is changing; it's just because one person moved on a different path, and therefore covered a different amount of spacetime. If one person walks in a straight line between two points, and another walks a crooked path between the same points, the second person travels a different distance, but we don't say that they experienced a different number of meters per meter.

When people talk about time slowing down in relativity, it's generally because they haven't truly, deep in their bones, let go of Newtonian absolute time. They're implicitly relating the time clicking by on

Bob's clock to some objective time out there in the universe. But there is no such objective or absolute time; there is only what clocks measure. We are welcome to put a time coordinate on spacetime, but that's far from objective. I might use a different coordinate system than you do, and nothing physical depends on this choice.

While Bob is zooming around in his spaceship, if he ever glances at his watch, it will appear to be ticking at precisely its normal rate, one second per second. After all, both Bob and the watch are moving along the same trajectory through spacetime, and therefore experiencing the same amount of proper time. If Bob takes his pulse, he will measure the same number of heartbeats per minute that he usually does. (Unless he's excited because he's zooming around in a spaceship, which would be understandable.)

But . . . you might be thinking—and here's the crucial bit—surely Bob's time has slowed down with respect to Alice's time, right? After all, when they reunite he has aged less.

The problem is that in order for us to make sense of the claim that Bob's clock ticks at a different rate from Alice's, we would have to somehow compare them. That's no trouble when the clocks are at the same location, where we can just look at them. But when they are separate, there's no way to "simultaneously" check what time it is on both clocks. If we're close to one clock and the other is far away, it will take time for any signal from the distant clock to reach us; no signal from there to here can move faster than the speed of light. By the time we see what the distant clock reads, more time has passed on the nearby clock.

You might think this is just a technical annoyance we could fix up by bouncing signals back and forth between clocks or some such thing. You are welcome to invent such schemes. But there's no way for the comparison to be absolute or instantaneous. You've just invented something arbitrary, not a natural way of comparing distant clocks to each other.

The right strategy is to give up on the idea of comparing clocks that are far away from each other. That's perfectly okay, and very much in the spirit of relativity. Think locally, and let go of the parochial impulse to extend your local thoughts into absolute structures across the span of space and time.

MINKOWSKI SPACETIME

You've been very patient with all these words. Time to get to some equations.

We've emphasized that the time measured on clocks moving through spacetime is analogous to the distance traveled along a path through space. We can put some quantitative meat on those bones by thinking about how we measure distances.

The answer lies in **Pythagoras's theorem**. Consider a right triangle, a triangle where one interior angle is a right angle. The long side, the hypotenuse, will be opposite the right angle. Then Pythagoras tells us that the square of the length of the hypotenuse is equal to the sum of the squares of the lengths of the other two sides: $a^2 + b^2 = c^2$, if a and b are the short sides and c is the hypotenuse.

That's an intrinsically interesting result, especially for fans of right triangles. Where it becomes crucial for us is when we express space in terms of Cartesian (perpendicular) coordinates. For simplicity let's just consider two-dimensional space and label our coordinates x and y as usual.

For any two points on the plane, there is a well-defined distance d between them. Given our coordinate system, there is a nice way to calculate what that distance is. We can draw a right triangle by starting at one point, traveling in the x-direction (horizontally) until we reach the x-value of the second point, then traveling in the y-direction (vertically) until we reach the second point itself. This defines a right triangle, whose shorter sides have lengths Δx and Δy, and the hypotenuse is the distance we're interested in. (Remember that the Greek letter Δ, Delta, means "the change in.") So by Pythagoras's theorem, the distance can be expressed in terms of the changes in coordinates between the two points as

$$d^2 = (\Delta x)^2 + (\Delta y)^2. \tag{6.1}$$

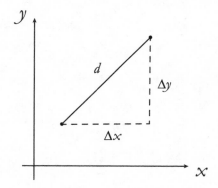

You probably knew all that. The cool part is that almost the same thing works for the time measured along a path in spacetime. "Almost" because there is one crucial modification—in spacetime, a minus sign sneaks into Pythagoras's theorem. It is this minus sign that is responsible for the switch from "shortest distance" to "longest time" for straight paths.

Consider a simplified two-dimensional spacetime on which we put coordinates x for space and t for time. And imagine we have a straight-line path (constant velocity) traveling between two events. Let τ (the Greek letter tau) be the proper time, what a clock would measure along the path. We can once again define the coordinate differences between the two events, this time Δx for space and Δt for time.

The defining feature of Minkowski spacetime, the arena on which

the drama of special relativity plays out, is that the proper time satisfies a Pythagoras-like equation, but with a crucial minus sign in front
of the spatial bit:

$$\tau^2 = \left(\Delta t\right)^2 - \left(\Delta x\right)^2. \tag{6.2}$$

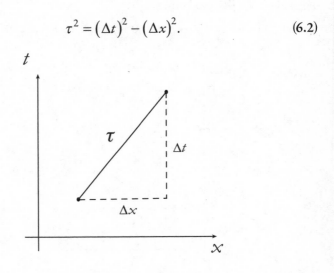

This simple equation tells us a lot about how spacetime works (everything we need to know, really). Think of a **stationary observer**, someone who doesn't move at all in the coordinate system we have set up.
They will age forward in time, of course, but they don't move in space.
So for a stationary observer we have $\tau^2 = \left(\Delta t\right)^2$, or simply $\tau = \Delta t$. The
proper time and the coordinate time are the same, which is what most
of us are used to anyway.

Observers who move tell a different story. They will have a nonzero Δx. So their proper time will always be *less* than that of a stationary observer, given a fixed amount of coordinate time elapsed, because
of that funny minus sign in (6.2). Relativity posits a trade-off between
time and space: On a journey between any two fixed events, the more
you move in space, the less time you feel passing.

We've been talking about straight-line paths, but there's no trouble
whatsoever with incorporating arbitrary journeys into our story. We

can guess what the strategy should be: Apply an infinitesimal version of (6.2) to very short segments of a path, then do calculus. For infinitesimal distances in spacetime, the modified Pythagoras formula becomes

$$d\tau^2 = dt^2 - dx^2. \qquad (6.3)$$

If we want to calculate the total amount of proper time elapsed along a path, we just integrate to get $\Delta\tau = \int d\tau$.

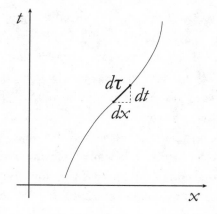

Sometimes people will suggest that special relativity works only for unaccelerated trajectories, and you need general relativity to handle acceleration. Rubbish. General relativity becomes important when spacetime is curved and we have gravity. As long as spacetime is flat—which it is in Minkowski spacetime, which we're sticking to in this chapter—special relativity applies, and you can consider any paths you like.

THE SPEED OF LIGHT

Confession time. We cheated a bit in the modified Pythagoras relations, equations (6.2) and (6.3). Remember dimensional analysis:

Physical quantities are expressed in units, and we can only add quantities together if they have the same units. But τ^2 and $(\Delta t)^2$ have units of (time)^2, while $(\Delta x)^2$ has units of (distance)^2. Who do we think we are, just adding them together like that?

The secret is that relativity provides a universal conversion factor between space and time. (They are both directions in spacetime, after all.) The conversion factor is denoted c and is numerically equal to

$$c = 299{,}792{,}458 \text{ meters/second.} \qquad (6.4)$$

You might know this quantity as **the speed of light**. But what's important is not that light travels at this speed. What matters is that there is a universal speed, built into the very fabric of spacetime, that we can use to convert between space and time. It just so happens that light waves (as well as gravitational waves, for that matter) travel at this special velocity.

This is such an important property that old-fashioned units like "meters" become cumbersome. It's immensely more convenient to use units like "light-seconds," the distance that light travels in one second. Namely, one light-second equals 299,792,458 meters. This number is exact, because it is easiest to define one second in terms of certain atomic vibrations, then define the meter to be the distance light travels in 1/299,792,458 seconds.

It might be exact, but that's an awkward number. If we use light-seconds instead of meters, we can simply write

$$c = 1 \text{ light-second/second.} \qquad (6.5)$$

That's just as exact, and much easier to remember. Things work equally well if we use light-years and years, or whatever.

The great thing about using units where $c = 1$ is that you can just leave the speed of light out of your equations entirely, since multiplying and dividing by 1 doesn't do anything. That's what we sneakily did when we wrote down (6.2). On the right-hand side, where we wrote Δx, what we really meant was $\Delta x/c$. Once we've chosen units in which $c=1$, we can take the speed of light as implicit. That's both notationally cleaner and helps reinforce the deep truth that distances and times are both measures of displacement in spacetime.

Now we can plug in numbers to the twin thought experiment. Consider the segment of Bob's journey from when he first departs to the midpoint when he turns around. (The second segment is a repeat of the first, just with the direction of travel reversed, so we don't have to do the calculation twice.) We are assuming he is moving at a constant velocity,

$$v = \frac{\Delta x}{\Delta t}. \tag{6.6}$$

Equivalently, we know that $\Delta x = v\Delta t$. Plugging into (6.2) gets us

$$\begin{aligned} \tau^2 &= \left(\Delta t\right)^2 - \left(\Delta x\right)^2 \\ &= \left(\Delta t\right)^2 - v^2\left(\Delta t\right)^2 \\ &= \left(1 - v^2\right)\left(\Delta t\right)^2, \end{aligned} \tag{6.7}$$

or

$$\tau = \sqrt{1 - v^2}\,\Delta t. \tag{6.8}$$

Alice and Bob start and end at the same events, traversing the same amount of coordinate time Δt. Alice is stationary, so her proper time is equal to the coordinate time elapsed. But Bob experiences less proper time, by a multiplicative amount $\sqrt{1-v^2}$ (in units where $c=1$, remember). If $v = 0.99$, we get $\sqrt{1-v^2} = 0.14$, and $1/0.14$ is approximately 7. That's why Alice ages seven years for every one of Bob's.

This calculation also explains why it took so long to invent relativity. Newton was smart—why did he ever think that you could group everyone's proper time into some absolute time? The answer is that most of us spend our lives moving much more slowly than the speed of light. If you're driving in a car at 65 miles per hour, that's $v = 10^{-7}$ in units where $c = 1$. In such cases, the factor $\sqrt{1-v^2}$ in (6.8) is approximately 0.999999999999995. That's so close to 1 that you'd never tell the difference. In our everyday lives, the proper times we experience are essentially indistinguishable from a single background coordinate time.

LIGHT CONES

The minus sign in the proper-time formula (6.2) opens up an interesting possibility. If we consider a straight path that traverses equal amounts of space and time, $(\Delta x)^2 = (\Delta t)^2$, we will have $\tau = 0$. So the object moves, but no proper time elapses along its journey. These are just trajectories that move at the speed of light:

$$v = \frac{\Delta x}{\Delta t} = \pm 1. \tag{6.9}$$

(This minus sign just means the object could move to the left rather than to the right, which doesn't matter for our present purposes.)

That's a pretty interesting result. Anything that moves at the speed of light, including light itself, doesn't experience the passage of time. It's tempting to ask yourself what it would be like to travel at that speed. (Apparently Einstein wondered about this when he was in high school.) The short answer is that there is nothing "it would be like" to travel at the speed of light; if somehow you could do it, you wouldn't experience the passage of time, so you wouldn't have any perceptions or conscious thoughts at all.

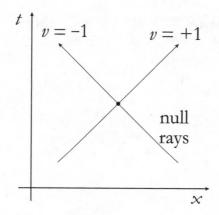

Paths in spacetime that move at the speed of light are called **null** trajectories (because the proper time is zero) or **lightlike** trajectories (for obvious reasons). When we choose units where $c = 1$, null trajectories are drawn as diagonal lines that tilt at 45 degrees on a spacetime diagram, traversing equal amounts of space and time. Anything that is moving slower than the speed of light will traverse more time than space, so such paths are known as **timelike** trajectories, moving upward in the spacetime diagram. We can also consider **spacelike** trajectories, which move sideways and traverse more space than time. No physical objects can travel on spacelike paths (you can't go faster than light), but that doesn't stop us from drawing such paths or thinking about them.

Pick a single event in spacetime and call it A. Imagine drawing all the lightlike rays starting at that event and extending into the future. It's as if a lightbulb flashed on and then immediately off again at that point, and we are tracing the paths of all the photons of light created in that instant. (The spacetime version of poking your finger into still water and watching a ripple travel outward, if that ripple moved at the speed of light.) Together, those paths constitute the **light cone** of event A. Our hypothetical lightbulb doesn't need to actually exist; every event defines a light cone of null rays connected to it, even if no light is traveling on those rays. In fact, there is a past light cone as well as a future one, consisting of all the lightlike paths pointing directly toward *A* from the past.

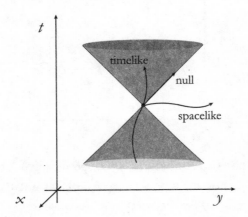

The speed-of-light limit implies that physically allowed trajectories that pass through any event must, as they stretch from past to future, stay inside the light cones of that event. This is true for every event; each point in spacetime defines past and future light cones, and any physical trajectory passing through them must stay inside.

This light-cone structure is what replaces the quaint Newtonian notion of absolute space and time. When we draw a Newtonian spacetime diagram, we slice spacetime into horizontal "moments of time." Two events that are separated in space but occur at the same time are said to be "simultaneous."

That's not the right way to think once we've made the transition to relativity. There is no notion of simultaneity between two separated events. There is only whether two events are inside each other's light cones. If they are, we say the events are "timelike separated," while if they're outside, they are "spacelike separated." Those are well-defined notions that every observer in the universe will agree on, while "simultaneous" is not.

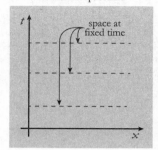

Newtonian spacetime

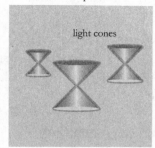

Minkowski spacetime

Slicing spacetime into horizontal moments seems natural because, once again, we live our lives moving very slowly compared to the speed of light. For everyday purposes, it's much more convenient to use units like "meters" rather than "light-seconds." (In all of human history, only a few Apollo astronauts have ever traveled farther than one light-second from their birthplace on Earth. And it took them much longer than one second to do so.) If we draw a spacetime diagram using meters and seconds, light cones aren't tilted at 45 degrees. Instead, they zip through so many meters in a single second that they look almost horizontal. It's hard to tell the difference between "the set of all spacelike-separated events" and "a single moment of time." But the difference is very real.

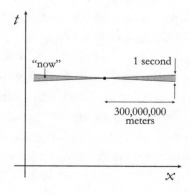

Once you become hard-core about relativity, you would be well served to not even draw "slices of space at fixed time" on your space-time diagrams. You should draw only light cones, which are objective and universal. The urge to slice spacetime into moments of time is almost irresistible. Just don't think you're reflecting any true features of reality.

REFERENCE FRAMES

Having said all that, let's acknowledge that people are going to keep drawing horizontal slices on their spacetime diagrams, even if they've been brought up to know better. These slices are called "reference frames," or "global reference frames" to emphasize that we are extending them throughout all of space. When we make the transition from the flat spacetime of special relativity to the curved spacetime of general relativity, global reference frames go from "unnecessary" to "maybe not even possible." But thinking about them helps us understand why people talk about ideas like "length contraction," so it's worth sorting them out, if only so we can better communicate with our fellow special-relativity enthusiasts.

Let's start as usual by reminding ourselves how things work in space. In two-dimensional flat space, it's often convenient to use Cartesian coordinates x and y, which define axes that are perpendicular to each other. But everyone knows that it's the two-dimensional plane itself that matters, not the particular coordinates we use to locate points within it. We can choose other coordinates x' and y', for example, by rotating the original ones by some angle. The new coordinates are just as useful as the old ones, and physically measurable quantities like the distance between two points don't change.

We can do a similar thing in spacetime. Consider a single observer who doesn't accelerate in any way and use them to set up a coordinate system. They have a clock with them, so we can use the proper time along their trajectory to define a time coordinate t. We also, a bit more

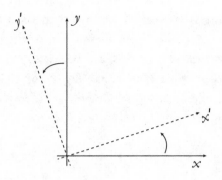

creatively, imagine that once every second our observer shoots out "infinitely fast" rays in all directions. We all know nothing can travel infinitely fast, so this is a purely conceptual move. What we really mean is that our observer defines spacelike lines that are "perpendicular" to their world line. Let the distance along these rays be labeled x. In this way we define a coordinate system—an **inertial reference frame**, since our observer is unaccelerated and hence "inertial"—that extends throughout spacetime.

You can probably guess what's coming next: We do the same thing, but now starting with a different observer, one who is also unaccelerated but who is moving at a constant velocity with respect to the first one. They define a time coordinate t' using their clock, and shoot out imaginary infinite-speed spacelike rays to define a spatial coordinate x'. Basing a new coordinate system on a moving observer is analogous to rotating a Cartesian coordinate system in ordinary space. This change of coordinates is called a **Lorentz transformation**, after Dutch physicist Hendrik Antoon Lorentz.*

But there is a surprise. When we plot the infinite-speed rays

---

* Lorentz's work was so central to special relativity that the theory was occasionally referred to as "Lorentz–Einstein theory" in the early days. But Lorentz himself continued to cling to the idea of a luminiferous aether that defined an absolute frame of rest.

emitted by the moving observer on our original spacetime diagram, they don't *look* perpendicular to the motion of the observer themselves. The new "time axis" and "space axis" (t', x') look like they are scissoring together rather than remaining at right angles. But they actually are perpendicular to each other; the difference can be traced to that minus sign in the Minkowski distance formula (6.2).

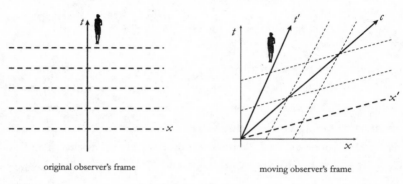

original observer's frame   moving observer's frame

From a physical perspective, this scissoring effect can be thought of as a consequence of the constancy of the speed of light. We notice that the new time and space axes maintain a constant angle with respect to the light cones. This reflects the fact that every observer will measure the speed of light to be equal, whatever reference frame they are using.

This is a good place to mention that the phrase "theory of relativity" is a misnomer. The basic meaning of "relativity" in this context is that there is no objective, preferred reference frame in the universe; we can only measure the velocity of an object relative to the motion of other objects, not in any absolute sense. But that was also true in Newtonian mechanics, which featured Galilean relativity. The "theory of relativity" in the modern sense comes from the combination of this relativity principle and the constancy of the speed of light to all observers. Or, equally well (and more elegantly), it comes from the idea that we live in Minkowski spacetime, where proper time is measured by (6.2).

LENGTH CONTRACTION

The tilting of reference frames helps explain the famous phenomenon of **length contraction**, according to which objects moving at high velocity supposedly get shorter. After all, what is the "length" of something? You might think that a ruler, for example, has a length. But unless your ruler literally exists only for a single moment, it has extent in time as well as in space. If we simplify our lives by abstracting a physical ruler to an object with just one dimension of spatial extent, it still has a two-dimensional "world volume" in spacetime. When we talk about its length, we need to choose a reference frame to distinguish the spatial part from the temporal part of the ruler's world volume. The "length of the ruler" is actually the spatial distance of a cross section of that world volume in a particular reference frame. Usually, with good reason, we consider the rest frame of the ruler, the coordinate system in which it's not moving. Talking about the length of a moving ruler is equivalent to measuring that spatial distance in a different frame.

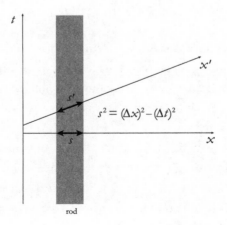

$$s^2 = (\Delta x)^2 - (\Delta t)^2$$

rod

We can see from the figure that it does indeed look like the spatial cross section of our ruler will be different in a moving reference frame. Unfortunately, it looks to be longer, rather than shorter, in the moving frame. That seems like elongation, not contraction. What's going on?

The culprit is our intuition about how length works when we draw a map of space, which doesn't always carry over to spacetime diagrams. The formula (6.2) tells us how to calculate the proper time along a trajectory in terms of time and space coordinate displacements. But "proper time" only makes sense for timelike trajectories. For spacelike trajectories, Δx is greater than Δt, which would mean that τ^2 would be negative, which is not a good quality for the square of a number to have. (It doesn't make sense to think of an interval in spacetime as an imaginary number.) There is no reason to attach a proper time to a spacelike interval.

What we want to attach to spacelike intervals is, of course, a spatial distance, which in this context is usually denoted s. Happily, there is a version of (6.2) that works well for this purpose: Just reverse the sign. Here is the formula for the length of a spacelike segment:

$$s^2 = \left(\Delta x\right)^2 - \left(\Delta t\right)^2. \tag{6.10}$$

The minus-sign difference between time and space explains why the length of the ruler as measured in the moving frame is actually shorter than in the rest frame, even though it looks longer on the page: because it extends in time as well as in space. In a two-dimensional spatial plane, picking up the end of a horizontal line segment and moving it vertically always makes the segment longer, because of Pythagoras's theorem. But in spacetime, moving one end of a segment a bit in time will make it shorter. Length contraction is real, but it's not that the object is physically shrinking; we're just in a different reference frame, measuring a different quantity.

SIMULTANEITY AND ITS DISCONTENTS

Comparing these two reference frames drives home the harsh reality that there is no such thing as "at the same time" for distant events in relativity. That's pretty obvious from the figure, since the lines defining "simultaneous" aren't the same in the two coordinate systems. But it gets worse.

Take the event where our two observers overlap, the origin of both coordinate systems, and label it A. Then choose another event B, far away from A in space and just a tiny bit in the future as measured in our original time coordinate t.

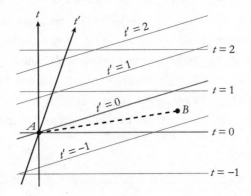

From the figure we can immediately see that while B might lie to the future of A in the (t, x) coordinate system, it's in the *past* of A in the (t', x') system.

That's just what life is like for spacelike-separated events. There is no such thing as which event is "really" in the past or the future. That depends on what reference frame we use, and all reference frames are equally valid. All that we can say in good conscience is that the two events are spacelike-separated.

It's natural for us to think about what's happening "right now" somewhere else in the world. That kind of thinking will get us into trouble when we start contemplating astronomically large distances.

The nearest star to our sun, Proxima Centauri, is approximately four light-years away. For any particular event on Earth, there is an eight-year span of events on Proxima Centauri that could count as "simultaneous" with it, depending on your reference frame.

This feature of special relativity spells trouble for *Star Trek* and other space operas. It's common in such stories to posit the invention of warp drive or some other advanced technology that allows our heroes to travel faster than light. But once you can travel on spacelike trajectories, you can effectively travel backward in time, at least in some reference frame. Indeed, all spacelike trajectories are created equal, according to special relativity. So if you can travel faster than light, you can go backward in time in any reference frame at all.

Maybe that's what you want for your science-fictional universe. (*Star Trek* has certainly been pretty shameless about invoking time travel when it seemed like fun and then ignoring it in subsequent episodes.) But it's hard to keep things logically consistent. It's safer for physics and for fiction to just exclude faster-than-light travel entirely.

UNIFICATION

James Clerk Maxwell's theory of electromagnetism was one of the first great examples of **unification** in physics, bringing together electrical and magnetic phenomena under a single umbrella. (Newtonian gravity was an even earlier case, unifying the falling of apples with the motion of the planets.) Numerous beneficial insights came along for the ride, such as the understanding that light is a form of electromagnetic wave. Special relativity, which was inspired by Maxwell's theory, is another great example, unifying space and time together into spacetime. We shouldn't be surprised that this perspective comes with its own new insights.

In three-dimensional space we often talk about vectors, such as the velocity vector of a moving object. As we know, that's just the derivative with respect to time of the object's position, $\vec{v} = d\vec{x}/dt$. Given some coordinate system in space, it's convenient to express the vector in terms of its components in each direction:

$$v^i = \left(v^x, v^y, v^z \right) = \left(v^1, v^2, v^3 \right). \tag{6.11}$$

Here the superscript i is an **index**—a label on which component we're talking about, not an exponent!—running over the numerical values $(1, 2, 3)$, corresponding to the spatial directions (x, y, z).* Note the flexibility of this notation: We can think of i as standing for either the numerical index of which component we're considering or the literal coordinate variable.

Now that we've progressed from space to spacetime, it's natural to define a generalization of these three-dimensional vectors, which we will imaginatively call **four-vectors**, since spacetime is four-dimensional. The spacetime version of velocity is the **four-velocity**, which we can think of as the rate at which the object moves through spacetime.

We will first invent some slick new notation that will look intimidating at first but will quickly become second nature. Namely, we will include time among our coordinates on spacetime by promoting the Latin index i, which runs over $(1, 2, 3)$, to a Greek index μ (mu), which will run over $(0, 1, 2, 3)$. That means we are treating time as the "zeroth coordinate," $t = x^0$. You might think it would be more natural to make time the fourth coordinate, but labeling it as x^0 will make our lives easier if we ever want to contemplate additional (or fewer) dimensions of space. The set of spacetime coordinates is therefore

---

* Superscripts do look a lot like exponents, but in this case it's just a label on the three components. You might wonder why we don't just use subscripts, but once we get to general relativity we will need to use both, and they will stand for slightly different things.

$$x^{\mu} = \left(t, x, y, z\right) = \left(x^0, x^1, x^2, x^3\right). \tag{6.12}$$

The four-velocity of an object is going to be the derivative of its coordinates in spacetime.

But wait a minute. You've already been forced to sit through a rant about how it doesn't make sense to talk about "the speed of time" in the same way we talk about speed through space. So how can we talk about a velocity through spacetime?

The main trick is to define the four-velocity as a derivative with respect to the proper time τ along the trajectory, rather than with respect to the coordinate time t. We write the components of the four-velocity as

$$V^{\mu} = \left(V^0, V^1, V^2, V^3\right) = \frac{dx^{\mu}}{d\tau} = \left(\frac{dt}{d\tau}, \frac{dx}{d\tau}, \frac{dy}{d\tau}, \frac{dz}{d\tau}\right). \tag{6.13}$$

What does the four-velocity physically represent? From (6.8) we know that

$$d\tau = \sqrt{1 - v^2}\, dt. \tag{6.14}$$

(We showed this for straight-line paths, but it's true generally for infinitesimals, with v being the velocity at the point under consideration.) So the components of the four-velocity become

$$V^{\mu} = \frac{1}{\sqrt{1 - v^2}} \left(1, \frac{dx}{dt}, \frac{dy}{dt}, \frac{dz}{dt}\right). \tag{6.15}$$

When v is very small, the factor $\sqrt{1 - v^2}$ is approximately equal to 1 and can be ignored. Then the zeroth component V^0 of the four-velocity is simply 1, and the other components are just equal to the ordinary three-velocity $\vec{v}$. Sometimes we will write this as $V^{\mu} \approx \left(1, \vec{v}\right)$ (valid for small velocities).

Now comes the real magic. In pre-relativistic Newtonian mechanics, momentum is represented by a three-vector given by mass times velocity, $\vec{p} = m\vec{v}$. In the world of relativity, we define the four-momentum in an analogous way, as mass times four-velocity:

$$p^{\mu} = \left(p^{0}, \, p^{x}, \, p^{y}, \, p^{z} \right) = mV^{\mu}. \tag{6.16}$$

Comparing to (6.15), we see that the spatial components here are quite similar to the Newtonian expression for the momentum, with an extra factor of $1/\sqrt{1-v^2}$. What about the time component? We have

$$p^{0} = \frac{m}{\sqrt{1-v^2}}. \tag{6.17}$$

So the time component of the four-momentum is just the mass of the object divided by a velocity-dependent factor.

As usual, we gain some intuition by considering small velocities (the "non-relativistic limit"). At this point let's invoke a powerful mathematical trick. Whenever we have an expression $\left(1 + x\right)^n$, where x is a small number and n is some fixed exponent, a good approximation is given by

$$\left(1 + x\right)^{n} \approx 1 + nx + \cdots \tag{6.18}$$

The dots indicate that there are more terms (an infinite number, when n is not a positive integer), but they will be proportional to x^2, x^3, and higher powers of x. Since we're assuming x is very small, those higher powers will be even smaller and can be ignored.

Our ubiquitous factor of $1/\sqrt{1-v^2}$ is precisely of this form, where $x = -v^2$ and $n = -1/2$. (Taking the square root of a quantity is the same as raising it to the power ½, and taking the reciprocal of a quantity is the same as raising it to the power of -1.) So for small velocities we get

$$\frac{1}{\sqrt{1-v^2}} \approx 1 + \frac{1}{2}v^2. \qquad (6.19)$$

Plugging this into (6.17), we find that

$$p^0 \approx m + \frac{1}{2}mv^2. \qquad (6.20)$$

Hmm, that second part looks familiar; it's the kinetic energy. Apparently the zeroth component of the four-momentum is an energylike thing that includes a constant term m plus the kinetic energy.

Here's an idea: Let's define the energy of an object in relativity to just be that zeroth component of the four-momentum:

$$E = p^0 = \frac{m}{\sqrt{1-v^2}}. \qquad (6.21)$$

As a side benefit, this expression helps us understand why we can never accelerate a rocket to go faster than the speed of light. As v gets closer and closer to 1, the quantity $\sqrt{1-v^2}$ approaches zero, and $E = m / \sqrt{1-v^2}$ goes to infinity. It would take an infinite amount of energy to accelerate a finite-mass object to the speed of light, much less beyond it. That's not going to happen.

When the velocity is much less than the speed of light, we can use (6.20) to write

$$E \approx m + \frac{1}{2}mv^2. \qquad (6.22)$$

We can think of the kinetic energy as "the energy of the object that is due to its motion," and the other term, which is just the mass m, as "the energy that the object has even when it is at rest." Call that the **rest energy**, $E_{\text{rest}} = m$.

The units aren't quite right, presumably because we've been setting $c = 1$. From the formula for kinetic energy we know that energy has units of mass times velocity squared. So we can fix the units by multi-plying by c^2. This leads us to a celebrated conclusion:

$$E_{\text{rest}} = mc^2. \tag{6.23}$$

You will most often see this without the "rest" subscript, but that's because we are being careful and other people are sometimes sloppy. The right way to think about this famous formula is that objects have energy even if they are sitting completely at rest, and that rest energy is equal to the mass times the speed of light squared. (Or you could think of what is meant by "the mass" as "the value of the four-momentum in the object's rest frame." Either way works.)

This is the most famous example of the conceptual unification provided by special relativity. Energy and momentum aren't two distinct ideas; *energy is the timelike version of momentum.* One of the wonderful features of physics is how disparate notions can be brought together by the power of a good theory.

SEVEN

GEOMETRY

When Hermann Minkowski suggested in 1907 that the best way to think about special relativity was in terms of a unified four-dimensional spacetime, Einstein was skeptical. He complained in print that Minkowski's approach "makes rather great demands on the reader in its mathematical aspects."

But Einstein soon grew to appreciate the spacetime formalism, especially as he turned his attention to the question of how to incorporate gravitation into relativity. He ultimately became convinced that gravity could be understood as a manifestation of the curvature of spacetime itself. That's a priceless insight, but it doesn't count as a real physical theory unless you can turn it into an appropriate set of equations. Such equations would come from geometry: in particular, **Riemannian geometry**, which allows spaces to be arbitrarily curved and studied from the inside, rather than requiring them to be embedded in some higher-dimensional space.

The problem was, Einstein didn't know anything about Riemannian geometry. Almost no physicist of the time did; the subject had only been initiated in the 1850s, and into the 1910s it hadn't found any special use within physics. Fortunately, one of Einstein's old classmates, Marcel Grossmann, had become a professor of mathematics and was an expert in the subject. Under Grossmann's tutelage, Einstein learned enough about geometry to formulate general relativity, his theory of gravity.

Now it's our turn. If Einstein had to put aside his other work to study up on Riemannian geometry, who are we to resist? The subject is sufficiently subtle and unfamiliar that we'll devote an entire chapter to it—there's little question that it qualifies as a Big Idea—and then we'll put it to use in physics in the next chapter.

EUCLIDEAN GEOMETRY

We all have memories, pleasant or otherwise, of geometry from high school. There were a lot of triangles and circles and other shapes. The kind of geometry we learned back then is associated with Euclid, an ancient mathematician who worked in Alexandria not long after Aristotle was teaching in Athens. It can be thought of as "tabletop" geometry—properties of lines and curves that we can imagine drawing on a flat, two-dimensional surface. (Though it's easy enough to generalize to three or more spatial dimensions.)

What makes Euclid so influential is not so much any particular results—theorems about properties of geometric objects—as it is the approach he pioneered. Indeed, **Euclidean geometry** encompasses a number of classic results, many of which predate Euclid's work:

- Pythagoras's theorem: the square of the length of the hypotenuse of a right triangle is equal to the sum of the squares of the other two sides.

- The angles inside a triangle add up to 180 degrees (π radians).

- The circumference of a circle is $2\pi r$, where r is the radius of the circle.

- The area of the interior of a circle (known as a "disk") is πr^2.

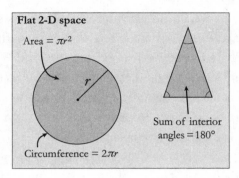

The distinctive feature of Euclidean geometry is that it defines an **axiomatic system**: We state a set of postulates, or axioms, and from them use the rules of logic to derive theorems. As long as you trust the basic rules of logic—which you should, although of course there are subtleties that philosophers care about—the theorems are as true as the axioms are. You don't necessarily need to take any stance on whether you think the axioms are true; a theorem is just a statement to the effect that "if these axioms are true, this result is too."

That's a very different workflow from what we have in science. Science is empirical and **fallibilistic**—any of our scientific theories could be wrong, no matter how much evidence we have so far accumulated for them. Scientists suggest hypotheses about how the world behaves, test them against data, and adjust our credence in them accordingly. You can never be absolutely sure that your hypothesis is correct, since data could come in tomorrow that changes your mind. But in geometry, and in mathematics and logic more generally, you can be sure that if your axioms are true, your theorems will be too.*

---

* From popular discussions, one might get the impression that there is a kind of hierarchy from a "hypothesis" (which is little more than a guess), up through "model" and "theory," and ultimately to "law," where the ideas are listed in increasing order of certainty. In the usage of real scientists, however, there is considerable overlap, to the point where there's no useful distinction between any of these

For the most part, Euclid's axioms are reasonable-sounding statements that make sense as a foundation for geometry. Things like "we can draw a straight line between any two points" and "all right angles are equal." But there was one that always stood out as different: Euclid's fifth postulate, known as the **parallel postulate**. One way of stating it is to construct two initially parallel lines in a plane, for example, by starting with a line segment and drawing two outgoing lines at 90 degrees to the segment. Then, according to the parallel postulate, those initially parallel lines will always remain an equal distance from each other. (Euclid's actual way of stating it was more like "if the angles are less than two right angles, the lines will eventually intersect.")

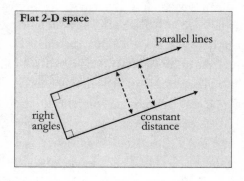

This also seems fairly reasonable, given our intuitions about how things work on a plane. (Mathematicians imagine planes that go forever in every direction, but we get pretty good intuition by thinking about tabletops or flat pieces of paper.) We do have to admit that it seems a bit more clunky than the others, however. For many years,

---

terms. There is a crucial and well-recognized difference, on the other hand, between "theory," which is a scientific model of how the world works, and "theorem," which is a rigorously proven mathematical result.

geometers thought it might be possible to prove the parallel postulate just using Euclid's other axioms, thereby moving it from the "axiom" side of the ledger over to the "theorem" side. When I took geometry in high school, our teacher trolled us by offering extra credit to anyone who could correctly prove it. None of us succeeded.

NON-EUCLIDEAN GEOMETRY

I will not play the same trick on you. You're not going to prove the parallel postulate from the other axioms of Euclidean geometry, because it's not possible to do so. We know that because you can replace the parallel postulate with a different postulate to add to the other axioms and obtain distinct versions of geometry, completely consistent in their own rights.

It's easy enough to see what the replacement postulates should look like. If the parallel postulate says that two initially parallel lines will maintain the same distance forever, the alternatives will presumably say that they *don't* maintain the same distance. And there are two ways that could happen: The lines could come together, or they could diverge.

Non-Euclidean Alternatives

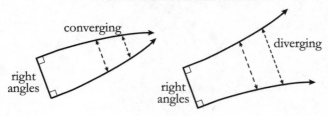

If you're worried that real straight lines don't behave that way, never fear. Our Euclidean intuition comes from our familiarity with the geometry of a plane or the equivalent. Planes are flat, not curved. These alternative postulates are supposed to hold for some kind of two-dimensional space, but a space with curvature, rather than the flat plane. Our alternative postulates define different kinds of geometries, cleverly named **non-Euclidean geometry**.

Which is not to say that the geometric systems based on these postulates are entirely abstract and hypothetical. There are, after all, two-dimensional shapes other than the plane, such as a sphere. (Whenever we talk about a sphere we just mean its surface, not the interior.) We can ask what would happen to two initially parallel lines on a sphere.

You might worry about what exactly we might mean by a "line," since nothing we draw on the surface of a sphere is going to appear perfectly straight. For the moment, just think about great circles or pieces thereof. A great circle is the curve we would find at the intersection of the sphere and a plane that passed through its center. The equator is an example, as are lines of longitude (not latitude!), but we can imagine great circles tilted at any angle.

So consider a segment of the equator of a sphere, and two lines going north at right angles. Let them be as straight as possible—that is, great circles. Like it or not, they are going to meet—in this case, at the north pole.

Sphere

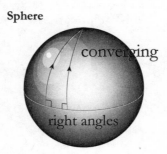

Other beloved features of Euclidean geometry also go wonky when we move from a plane to a sphere. Think of a circle of radius r, centered at the north pole. We can see from the picture that the circumference of this circle is *smaller* than $2\pi r$, and the area of the enclosed disk is smaller than πr^2. (If our radius goes all the way to the south pole, the circumference would be zero.) Meanwhile, the angles inside a triangle will generally add up to *greater* than 180 degrees. Indeed, we can construct a triangle with three 90-degree angles inside, by joining segments that go one-quarter the way around great circles. Euclid must be rolling in his grave.

Sphere

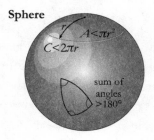

What about the case where initially parallel lines eventually diverge, rather than coming together? That one is harder to visualize, but it's similar to the shape described by a horse's saddle, or a Pringles potato chip.

Saddle

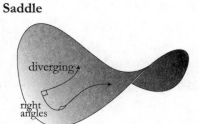

On a saddle-shaped geometry, properties of circles and triangles are once again modified, but in the opposite direction. For a circle with radius r, the circumference is going to be greater than $2\pi r$, the area of the enclosed disk is greater than πr^2, and the angles inside a triangle will generally add up to less than 180 degrees.

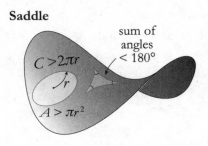

Saddle

In each of these cases there is an additional simplifying assumption lurking in the background: that whatever the geometric properties of our two-dimensional space are, they are exactly the same everywhere, and in every direction. If we start with a line segment of fixed length and extend initially parallel lines perpendicular to it, they will diverge/converge by the same amount, no matter where we stuck our segment or what direction it pointed in. The technical way of expressing this is to say that we are considering geometries of **constant curvature**, as opposed to curvature that changes from place to place or direction to direction. This makes things much easier, which of course means that we're going to unmake this assumption soon. It provides a natural starting point for a ride that will get wild quickly enough.

Once we're assuming constant curvature, there are only these three choices for the geometry of a two-dimensional space: Initially parallel lines stay parallel, converge, or diverge.

- Stay parallel: Euclidean geometry. Flat (zero curvature).

- Converge: Spherical (or elliptical) geometry. Positive curvature.

- Diverge: Hyperbolic geometry. Negative curvature.

INTRINSIC AND EXTRINSIC

The idea of non-Euclidean geometry didn't come onto the scene until the early nineteenth century, more than two millennia after Euclidean geometry. Nikolai Ivanovich Lobachevsky in Russia and János Bolyai in Hungary independently developed the basic ideas of hyperbolic geometry. It might seem puzzling, both that hyperbolic geometry was invented before spherical, and that it took so long in the first place. How hard is it to imagine drawing geometric shapes on a sphere, anyway?

These two questions—why did it take so long, and why did mathematicians stumble across hyperbolic geometry first?—in part answer each other. Of course everyone knew about spheres and had a basic idea of how lines and circles and angles worked on them. But they didn't imagine spheres as defining a separate kind of geometry. The two-dimensional spherical surface was thought of as **embedded** in three-dimensional Euclidean space. Indeed, one way of defining a sphere is "the set of all points at some distance R from a fixed point in space." Nobody was motivated to invent a whole new kind of geometry to describe spheres. Their properties could be derived from good old Euclidean geometry in three dimensions, applied to this particular embedded shape.

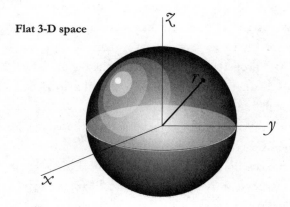

Flat 3-D space

Such is not the case for negatively curved surfaces. We can conceive of saddles or potato chips, of course, but close examination reveals that while these real-world examples have negative curvature, that curvature is not constant everywhere. There is a definite central point, and the curvature gradually declines as we move away from that point. No matter how clever we are, it's impossible to faithfully embed a two-dimensional space of constant negative curvature inside three-dimensional Euclidean space.

That's why the invention of hyperbolic geometry was a truly impressive intellectual achievement. We can write down a set of axioms that imply the geometric properties of two-dimensional space of constant negative curvature, sometimes called the **hyperbolic plane** (in contrast to the "flat" plane of Euclid). We can prove theorems within that system, derive formulas for the circumference of a circle and the area it encloses, and answer whatever other geometric questions we might pose. We just can't *build* such a space as a two-dimensional surface embedded in the space where we live. An exact hyperbolic plane exists only in our minds.

That's a big conceptual leap, as far as the history of mathematics is concerned. Freed from the self-imposed limitation of considering only objects we can actually build, mathematicians could study the implications of all sorts of axiomatic systems for their own sakes.

For our current purposes, the major ramification is the idea that we can study the **intrinsic** geometric properties of a space, as opposed to the **extrinsic** properties it inherits from being embedded in a larger space. This distinction was developed by Carl Friedrich Gauss, one of the greatest mathematicians of all time. (Gauss, who was famously slow to write up his results, also claimed to have invented hyperbolic geometry before Lobachevsky and Bolyai. But he never published anything about it, and you don't get credit for good ideas you keep to yourself.)

When we look at shapes drawn on a tabletop, or a sphere in three-dimensional space, we naturally perceive it from the outside. "Curvature," from that perspective, tells us how the shape bends and twists within the bigger space in which it's embedded. That's extrinsic curvature.

But we can also imagine how things would be perceived by imaginary beings that lived inside the shape. They could draw a circle and measure its circumference, for example. The result has nothing to do with how the shape is embedded in a bigger space; there doesn't even need to be any such space. The features that can be measured purely from the inside define the intrinsic geometry of the space.

Embedded curved geometry

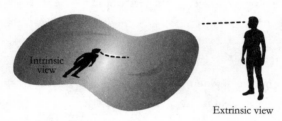

Intrinsic view

Extrinsic view

For a dramatic example of the distinction between intrinsic and extrinsic curvature, consider a two-dimensional cylinder embedded in three-dimensional space.

Cylinder

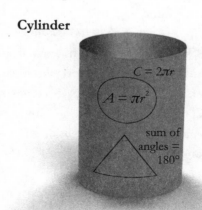

$C = 2\pi r$

$A = \pi r^2$

sum of angles = 180°

A cylinder sure *looks* curved, but that's an artifact of our external perspective. What we're perceiving is the extrinsic curvature, but intrinsically the cylinder is perfectly flat. You can check that by imagining drawing initially parallel lines and seeing whether they converge or diverge, or by drawing a triangle and adding up the interior angles, or by verifying the relationship between the radius of a circle and its circumference and area. In every case, the intrinsic geometry of the cylinder is the same as that of a flat plane.

We're dwelling on this because ultimately we're going to want to talk about the curvature of spacetime itself. Our universe isn't embedded in some bigger space; at least, it doesn't have to be, as far as we know. When it comes to spacetime, intrinsic geometry is all we have.

MANIFOLDS

Besides appreciating the distinction between intrinsic and extrinsic curvature, Gauss also pioneered the study of cases for which the curvature is *not* constant—spaces where curvature varies in amount and shape from place to place. But when it came to developing the full-blown theory of such arbitrary geometries, he handed the work off to his student Bernhard Riemann.

In 1853, Riemann was at the stage in his studies where it was time to give a lecture to complete his *Habilitation*, a qualification in the German system that is one step beyond the doctoral degree and is often a requirement to be allowed to teach university students. (For his doctoral work Riemann had pioneered the use of complex numbers to study two-dimensional surfaces, and the bulk of his *Habilitation* research was devoted to putting integral calculus on a rigorous footing. He was an ambitious and productive young man.)

The story goes that Riemann sent a list of possible lecture topics to Gauss, who surprised his student by picking the topic Riemann thought was the least interesting of the lot: the foundations of geometry. The assignment forced Riemann to sit back and think hard about what we really meant by "the geometry" of a space, and indeed what kind of "space" we are talking about, especially from the intrinsic insider's view. (Early in his lecture he complains that he was "not practiced in such undertakings of a philosophical nature," but he did pretty well at them.) He ended up writing one of the most influential papers in the history of mathematics, the results of which still sit at the heart of general relativity and the modern view of spacetime. Unfortunately Riemann died early of tuberculosis, or he may have derived any number of further foundational results in mathematics.

Riemann started his discussion by defining the notion of a **manifold**—an infinite set of points smoothly connected into a space of some definite dimensionality. Remember that when we zoom in closely, every curve looks like a straight line. Manifolds are constructed from the same kind of insight, but in more than one dimension. In a curved space, if we zoom in sufficiently close, things look like Euclidean geometry. The curvature that manifests itself on larger scales can be encoded in how infinitesimal bits of flat space are sewn together globally.

0-dimensional
manifold

1-dimensional
manifold

2-dimensional
manifold

The crucial thing about manifolds is that they don't need to be embedded in any other space, even if we sometimes draw them in ways that suggest they are. When we portray two-dimensional manifolds like a sphere or a torus, it's the two-dimensional surface we are thinking about, not the three-dimensional space in which they are embedded. That's just an artifact of how we three-dimensional creatures represent things. Manifolds have a well-defined topology and geometry in their own right—always think intrinsically, not extrinsically.

GENERALIZING PYTHAGORAS

Next, we have to decide how to specify the geometry of a manifold, once again referring only to intrinsic notions that could be measured by beings living inside. There are potentially a large number of ways to do this. The goal is to stipulate some basic geometric quantities of the manifold, from which anything else we might want to calculate can be derived.

The quantities that Riemann settled on were the lengths of curves. Not just one special curve: any curve we might contemplate drawing within the manifold. If you know the length of every possible curve you can draw, from that you can figure out anything you might want to know about the geometry.

It seems like an intimidating task, to specify the length of every possible curve. Even on simple manifolds like the sphere or the plane, there are an awful lot of curves we can draw. Happily, we have a tool for dealing with this unwieldy menagerie: calculus. We don't need to literally write down the length of every curve; we just need to give a formula for calculating the length of an infinitesimally small piece of a curve. Then we can get the entire length by integrating that infinitesimal piece.

To work up some intuition, consider a flat two-dimensional plane with Cartesian coordinates (x, y). Someone draws a curve of some sort on the plane, and we want to calculate the length ds of an infinitesimal piece of that curve in terms of infinitesimal displacements in the coordinates, dx and dy. We already know the formula for that: It's just Pythagoras's theorem. [If the notation here is giving you flashbacks to equation (6.10) for the length of a spacelike segment in Minkowski spacetime, that's entirely intentional.]

$$ds^2 = dx^2 + dy^2. \tag{7.1}$$

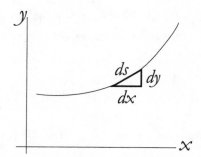

That's straightforward enough. But Cartesian coordinates are very special. Lines of constant x are all perfectly straight and always parallel to each other, as are lines of constant y. If the parallel postulate doesn't hold—if your manifold is not Euclidean—you can't draw coordinates like that everywhere. There are no Cartesian coordinates that cover a sphere, for example.

Even when our manifold is flat, we aren't forced to use Cartesian coordinates. Think about **polar coordinates** on a plane, where we specify a point in terms of its distance r from the origin and its angle θ from the horizontal axis. Consider the length of an infinitesimal segment in terms of these coordinates.

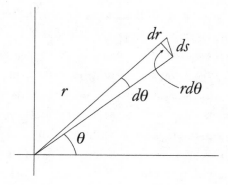

The physical length corresponding to an angular change $d\theta$ isn't a fixed amount; it grows larger as r increases. The length of a tiny segment at fixed r is going to be not $d\theta$ but $rd\theta$. That's straightforward enough to accommodate. The correct formula for an arbitrary infinitesimal segment is

$$ds^2 = dr^2 + r^2 d\theta^2. \qquad (7.2)$$

That's almost Pythagoras, but not quite. There's an extra factor of r^2 in front of the $d\theta^2$ piece. That extra factor encodes the idea that an increment $d\theta$ corresponds to more and more physical length as r increases. Essentially we are keeping the spirit of Pythagoras, but allowing for a slightly more general formula to relate physical distances to increments in coordinate values.

THE LINE ELEMENT

With that example as inspiration, what we want to understand is an idea called the **line element**—a general formula that can take an infinitesimal segment, characterized by infinitesimal displacements in each of the coordinates, and use those displacements to calculate the length ds of that segment. Let's start by sticking with two-dimensional manifolds for simplicity.

Imagine we have two coordinates (x^1, x^2), where these superscripts are indices, not exponents. (Just like we did for components of vectors.) We want to relate the distance ds to the coordinate displacements dx^i. Following our Pythagorean inspiration, we expect we should relate the squared distance ds^2 to squares of the coordinate displacements, like $(dx^1)^2$—that's the square of dx^1, so we're forced to use superscripts as both an index and an exponent in the same expression. And to be perfectly general we should allow for "cross terms" that multiply separate coordinates, like $dx^1 dx^2$. Finally, as we learned from polar coordinates, there can be coefficients multiplying these terms, which themselves can depend on the coordinates.

The most general such formula for the two-dimensional line element takes the form

$$ds^2 = A\left(x^1, x^2\right)\left(dx^1\right)^2 + B\left(x^1, x^2\right)dx^1 dx^2 + C\left(x^1, x^2\right)\left(dx^2\right)^2. \quad (7.3)$$

That's a lot of parentheses and superscripts, so let's take a breath and look at what's going on. The three quantities A, B, and C are numbers that take on specific values at each point on the manifold, so we have written them as functions of the coordinates (x^1, x^2). Each of them multiplies a product of two of the coordinate displacements dx^1 and dx^2. We have not only a term proportional to $(dx^1)^2$ and one proportional to $(dx^2)^2$, but also a new term proportional to $dx^1 dx^2$. That will be relevant if the coordinates are not perpendicular.

Here's why this expression is so important: Once you have told me the three functions $A(x^1, x^2)$, $B(x^1, x^2)$, and $C(x^1, x^2)$, the formula (7.3) gives us a way to calculate the length of *any* curve we draw. And that information, says Riemann, is sufficient to completely determine the geometry of the manifold. Everything we ever want to know—angles, areas, curvature—is contained in just those three functions. A similar story will hold in more than two dimensions, although the number of functions will be somewhat larger. In a d-dimensional space, it takes $d(d+1)/2$ functions to fully specify the line element.

Not that it's going to be simple to extract the geometric information we care about from these functions. The problem is that the line element can look very different in different coordinate systems, even if the underlying geometry is exactly the same. We've already seen this in the case of the flat plane. In Cartesian coordinates, the line element (7.1) is of the form (7.3), with

$$A\left(x, y\right) = 1, \; B\left(x, y\right) = 0, \; C\left(x, y\right) = 1. \quad (7.4)$$

While in polar coordinates, the line element (7.2) corresponds to

$$A(r, \theta)=1, \; B(r, \theta)=0, \; C(r, \theta)=r^2. \qquad (7.5)$$

Same geometry, but different versions of the line element, because we're in different coordinate systems. Geometry doesn't care about coordinate systems; those are human contrivances, not intrinsic features of the manifold. We're going to have to work just a bit harder to squeeze information about curvature out of these line elements.

THE METRIC

We're also going to have to invent some slick new notation. Good notation is its own reward, but the number of independent functions needed to specify the line element is going to quickly become unwieldy in higher numbers of dimensions.

This shouldn't be too hard, because we see the pattern from (7.3): For every pair of coordinate displacements dx^i and dx^j (where i and j are indices that run over the number of dimensions, and may be the same values or different), we assign a function of spacetime. We can label these functions as $g_{ij}(x)$, where now the unadorned variable x stands for all of the coordinates at once. The letters i and j have no meaning in and of themselves; they are just labels for the values of the indices, we can use whatever letters we like. So g_{11} is the function multiplying $(dx^1)^2$ in the line element, g_{12} is the function multiplying $dx^1 dx^2$, and so on.

These functions can be written in the form of a matrix,* for example, in three dimensions:

$$g_{ij} = \begin{pmatrix} g_{11} & g_{12} & g_{13} \\ g_{21} & g_{22} & g_{23} \\ g_{31} & g_{32} & g_{33} \end{pmatrix}. \qquad (7.6)$$

---

* We're using "matrix" in the mathematical sense of "an array of quantities." Nothing to do with whether or not we are living in a computer simulation.

This is the famous **metric tensor**, the object that will be at the center of our attention when we turn to general relativity. Each entry in the matrix, known as a **component** of the metric, is a separate function of position. Knowing all of those components tells us everything we need to know about the geometry of the manifold under consideration. Typically when a physicist is doing some research problem in general relativity, they are either trying to figure out what the metric is (for example, in the presence of a certain distribution of matter and energy), or working out the physical consequences of some particular form of the metric. (In relativity our metric will be on spacetime, not just space, but that turns out to make surprisingly little difference to the mathematical formalism. We'll just use Greek letters to label components.)

Knowing the metric is precisely equivalent to knowing the line element. The relationship is simply

$$ds^2 = \sum_{i,j} g_{ij} dx^i dx^j$$

$$= g_{11} \left(dx^1 \right)^2 + g_{12} dx^1 dx^2 + g_{13} dx^1 dx^3 + \cdots$$

(7.7)

So if we know all the components of the metric, we can calculate the length of any curve. From there, although it may not be obvious, we can also calculate areas and volumes, angles between lines, and much more.

Our two-dimensional line element (7.3) fits into this pattern, as it must. In this case we have

$$g_{ij} = \begin{pmatrix} g_{11} & g_{12} \\ g_{21} & g_{22} \end{pmatrix} = \begin{pmatrix} A & \frac{1}{2}B \\ \frac{1}{2}B & C \end{pmatrix}.$$

(7.8)

We see that B appears twice but is divided in half both times. That's because a literal application of the formula (7.7) gives us separate contributions for $dx^1 dx^2$ and $dx^2 dx^1$, but those are equal to each other.

We therefore require that the metric be symmetric: For each i and j we have $g_{ij} = g_{ji}$. From the point of view of the matrix version, components in the upper right will be equal to their corresponding components in the lower left.

Let's bring this down to Earth with some examples. We can start with familiar territory: flat three-dimensional Euclidean space in Cartesian coordinates (x, y, z). There we know what the line element is; it just comes from Pythagoras's theorem.

$$ds^2 = dx^2 + dy^2 + dz^2. \qquad (7.9)$$

The metric tensor, written as a matrix, couldn't be simpler:

$$g_{ij} = \begin{pmatrix} 1 & 0 & 0 \\ 0 & 1 & 0 \\ 0 & 0 & 1 \end{pmatrix}. \qquad (7.10)$$

(Henceforth we assume you recognize that the upper left component is g_{11}, the one to the right of that in the upper row is g_{12}, and so on, so we won't write that notation explicitly.) This is the **Euclidean metric**, which can be extended to any other number of dimensions by adding more rows and columns, keeping 1's on the diagonal and 0's everywhere else. Euclid himself was implicitly using it all the time, though he wasn't thinking in those terms. The metric defines the geometry.

But there are also non-Cartesian coordinate systems. For polar coordinates in two dimensions (r, θ), we can read off the answer from (7.2):

$$g_{ij} = \begin{pmatrix} 1 & 0 \\ 0 & r^2 \end{pmatrix}. \qquad (7.11)$$

It should be clear that the components of a metric are always relative to a certain coordinate system. An expression like (7.11) is only meaningful if you know that we are using (r, θ) coordinates, with $x^1 = r$ and $x^2 = \theta$.

There is also the three-dimensional version of polar coordinates, the spherical coordinates (r, θ, ϕ) we mentioned in Chapter 4. The metric for flat space in these coordinates is

$$g_{ij} = \begin{pmatrix} 1 & 0 & 0 \\ 0 & r^2 & 0 \\ 0 & 0 & r^2 \sin^2 \theta \end{pmatrix}. \tag{7.12}$$

As the angle θ goes from zero at the north pole to 90 degrees at the equator then to 180 degrees at the south pole, $\sin\theta$ goes from 0 to 1 then back to 0. Its appearance in the g_{33} component, bottom right corner of (7.12), reflects the fact that the physical distance associated with a change in $x^3 = \phi$ is small near the poles and larger near the equator.

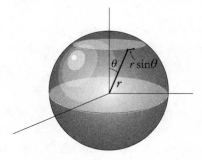

It cannot be overstressed that the geometry defined by (7.12) is the same geometry as that defined by (7.10). Both are "the Euclidean metric"—flat space—but one is expressed in Cartesian coordinates and the other in spherical coordinates.

THE METRIC ON CURVED SPACES
Let's look at a manifold with some curvature. The easiest example of that is the two-dimensional sphere. Happily, it's pretty easy to guess

the metric on a sphere once we've written down the metric (7.12) for flat space in spherical coordinates. A sphere is just the subset of flat space that we get from fixing the radial coordinate r to be some specific value R. So we can get the metric on a sphere by starting with (7.12), erasing the top row and leftmost column (since those tell us about distances in the r direction, which don't exist when we're confined to the sphere), and setting $r = R$. The result is the metric on a sphere in $(x^1, x^2) = (\theta, \phi)$ coordinates:

$$g_{ij} = \begin{pmatrix} R^2 & 0 \\ 0 & R^2 \sin^2 \theta \end{pmatrix}. \tag{7.13}$$

This bears a family resemblance to (7.12), the three-dimensional Euclidean metric in polar coordinates, but it's a very different beast. Here R is a fixed parameter telling us how big our sphere is, not a coordinate. And most important, this metric is not flat. It's a bit challenging to simply look at a metric and figure out whether it's flat or curved. Different choices of coordinates can obscure the underlying geometry.

Let's turn to the metric for the two-dimensional hyperbolic plane, as investigated by Lobachevsky and Bolyai (although they didn't call it that any more than Euclid did). As always, the first question is what coordinate system to use. A convenient one is known as the **Poincaré disk**. (It was first used by Eugenio Beltrami and only later by Henri Poincaré, but Poincaré is more famous, and famous people tend to get their names attached to things.)

The coordinates are generally labeled $(x^1, x^2) = (x, y)$, which makes us think of Cartesian coordinates on a plane, but there is an additional restriction that $r < 1$, where r is a radial coordinate measured from the origin. In this coordinate system the metric on two-dimensional hyperbolic space is

$$g_{ij} = \begin{pmatrix} \dfrac{4}{\left(1-x^2-y^2\right)^2} & 0 \\[2em] 0 & \dfrac{4}{\left(1-x^2-y^2\right)^2} \end{pmatrix} = \begin{pmatrix} \dfrac{4}{\left(1-r^2\right)^2} & 0 \\[2em] 0 & \dfrac{4}{\left(1-r^2\right)^2} \end{pmatrix}.$$

$$(7.14)$$

There are a few things going on here. The actual coordinates are (x, y), but in the second expression we've written the metric components in terms of $r = \sqrt{x^2 + y^2}$. That's no problem, you should just remember that r is a function of x and y. More significantly, the hyperbolic plane is infinite in size—it's a saddle shape extending infinitely far in all directions—but the coordinates we've used reach out to a finite value of r. That's perfectly okay. Remember all the way back in Chapter 2 where we noted that we can map an infinite length to a finite interval and back. That's what's going on here. The Poincaré disk *is* infinite in size, we've just covered it with coordinates that extend only over a finite interval.

In fact, you can tell that the space really is infinitely big, just from staring at the metric (7.14). Think about what happens as we approach the edge of the disk, $r \to 1$. Both of the nonzero metric components contain a factor $1/(1-r^2)^2$. As r gets closer and closer to 1, $(1-r^2)$ gets closer and closer to 0, so $1/(1-r^2)^2$ blows up to infinity. Physically, that means that any given coordinate increment dx or dy corresponds to a bigger and bigger actual distance. Even though the coordinates only reach over a finite interval, the physical size of the manifold being described is infinitely big. That's the magic of the metric at work.

The figure above shows a representation of hyperbolic space in Poincaré-disk coordinates, tessellated by triangles. The triangles look like they get smaller and smaller as they approach the edge, but that's an illusion caused by our coordinate system. In the metric (7.14), all of these triangles have the same size, and for that matter the same shape. There are an infinite number of them, squeezed down near the disk's edge. Dutch artist M. C. Escher famously made a series of engravings, which he dubbed *Circle Limit*, based on the geometry portrayed in this figure.

TENSORS

The metric on a manifold tells us how to calculate distances. We referred to it as the metric "tensor" but didn't explain what a tensor was more generally. What's up with that?

The idea of a *function* on a manifold is simple enough. It's a map from points on the manifold to the real numbers: an assignment of a number (the value of the function at that point) to each point. We might use a function to answer a question like "What is the density of matter at each point?" We're also familiar with *vectors*, which have both a magnitude and a direction. Vectors answer questions like "What is the velocity of a particle on this particular trajectory?"

Sometimes we want to answer more complicated questions, which may involve multiple vectors or directions in space. Maybe we want to know "how much overlap is there between vectors $\vec{v}$ and $\vec{w}$?" Or "how would a set of initially parallel trajectories twist around as they move through curved space?" **Tensors** are geometric quantities that contain the requisite information to address these more involved questions. Functions and vectors are kinds of tensors, but to think productively about curved spacetime we're going to need some more elaborate versions.

There are two ways to think about tensors, both of which are useful in different circumstances. One is the way we've already been using: as arrays of components, with each component labeled by a set of indices. All of those matrix presentations of the metric are examples of tensors in this sense. There are fussy requirements about how the components change when we change coordinates, but we don't have to worry about those.

The components of a square matrix have two indices (one indicating which row, another indicating which column). But tensors don't have to be square matrices. They can have any number of indices, and the values of the indices always run over the number of dimensions of the manifold we are thinking about.*

A vector is just a tensor with one index. We think of a vector $\vec{v}$ as an arrow with a length and a direction, but given a coordinate system such as (x, y, z) we can express the vector in terms of its components, which are the projections of the vector along each axis:

---

* The metric and other tensors we will care about can have different values at each point in space (or spacetime, when we get to that). So what we're really dealing with are tensor fields. A "scalar field" is just a function, taking on a numerical value at each point; we also have "vector fields" and other tensor fields.

$$v^i = \begin{pmatrix} v^x \\ v^y \\ v^z \end{pmatrix}. \tag{7.15}$$

So a vector is just a tensor with a single index, while the metric is a tensor with two indices. An ordinary function is a tensor with zero indices. And there can be more indices, which makes the tensor hard to represent as an array of components, but it can be done if you really try. For example, you can think of a three-index tensor as a vector of two-index tensors:

$$T^{ijk} = \begin{pmatrix} T^{1\,jk} \\ T^{2\,jk} \\ T^{3\,jk} \end{pmatrix} = \begin{pmatrix} \begin{pmatrix} T^{111} & T^{112} & T^{113} \\ T^{121} & T^{122} & T^{123} \\ T^{131} & T^{132} & T^{133} \end{pmatrix} \\ \begin{pmatrix} T^{211} & T^{212} & T^{213} \\ T^{221} & T^{222} & T^{223} \\ T^{231} & T^{232} & T^{233} \end{pmatrix} \\ \begin{pmatrix} T^{311} & T^{312} & T^{313} \\ T^{321} & T^{322} & T^{323} \\ T^{331} & T^{332} & T^{333} \end{pmatrix} \end{pmatrix} \tag{7.16}$$

I don't know why you would want to do that, but you're allowed. Once you have more than two indices, it's easier just to think about the individual components rather than writing the tensor as some big array.

The other way of thinking about tensors is as a map from one set of tensors to another tensor. Yes, that sounds circular, but it all hangs together in the end. For example, if you have two vectors v^i and w^j, you can make a number out of them using the metric. So you can think of the metric tensor as being a black box; you put two vectors in, and a number comes out.

The number in question is obtained by matching up indices on the metric with indices on the vectors, then summing over every index that is repeated both upstairs and downstairs.

$$g(v, w) = \sum_{ij} g_{ij} v^i w^j$$
$$= g_{11} v^1 w^1 + g_{12} v^1 w^2 + g_{21} v^2 w^1 + \cdots \qquad (7.17)$$

This is a famous construction, at least among people who spend their time pushing vectors around: It's the **inner product**, or "dot product," between the two vectors:

$$\vec{v} \cdot \vec{w} = \sum_{ij} g_{ij} v^i w^j. \qquad (7.18)$$

In ordinary Euclidean space, the dot product of two vectors is the product of the length of both vectors times the cosine of the angle between them. When the vectors point in the same direction, the dot product is just the product of their lengths, and when they are perpendicular the dot product is always zero.

A tiny secret is thereby revealed: The metric doesn't tell us only the length along curves, it also defines what we mean by "perpendicular." Two lines intersecting at a point are perpendicular when the inner product of two vectors pointing along them at that point vanishes. We can begin to see how the metric is determining everything we need to know about the geometry of a space.

You may have noticed that sometimes we write indices as super-scripts (as on vectors and coordinates) and sometimes as subscripts (as on the metric). This is not a whimsical choice; superscript and sub-script indices are distinct in important ways. All we need to know at the moment is that when we sum over indices, as in (7.17), we are only allowed to do so when one of the summed indices is upstairs and the other is downstairs. Any index that is summed over is called a **dummy index**, while one that is not summed over is a **free index**. Free indices can take any value—as long as the same index takes the same value in every term of an equation—while dummy indices don't have a "value." They are just shorthand for "add up every possible value of this index."

Summing over dummy indices happens so often in tensor analysis that Albert Einstein invented a helpful trick, known as the **Einstein summation convention**. Whenever we have a tensor or product of tensors where one or more indices are repeated both upstairs and downstairs, we can leave out the explicit summation sign and just as-sume they are being summed over. Thus we have, for example,

$$g_{ij}v^i w^j = \sum_{ij} g_{ij}v^i w^j. \tag{7.19}$$

Upon coming up with this innovation, Einstein joked to a friend, "I have made a great discovery in mathematics!" Summing over repeated indices isn't a deep concept, but adopting this helpful convention does save a great deal of time in a subject like general relativity.

PARALLEL TRANSPORT

Riemann's great insight was that the metric on a manifold contains all the information you need to extract its curvature and any other geometric information you might be interested in. Let's think how we can go about extracting that information. Our first step will be to push vectors around from place to place, because how they travel will be affected by the underlying curvature. For the rest of this chapter the mathematical machinery becomes cumbersome, so we're going to hit some highlights while relegating details to Appendix B.

Imagine you are standing at some location within a curved manifold, and you are holding a vector. Maybe you have a spinning gyroscope whose axis points in some spatial direction. And you have a friend some distance away who also has a vector. You would like to compare these two vectors: Are they pointing in the same or different directions, is one longer than the other, and so on. How can we do that?

In flat space—where most of our intuition comes from—this seems like a silly question. You could just walk over to your friend, carrying your vector along with you while keeping it fixed, and compare when you get there. But what does it mean, "keeping it fixed"? One answer is to set up a traditional Cartesian coordinate system and just keep all of the components of our vector constant. That allows us to move it from place to place, no problem.

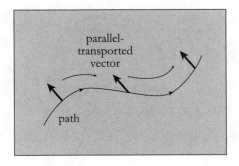

The problem is that this procedure doesn't work in non-flat geometries. There are no "Cartesian coordinates," because the existence of such coordinates relies on a flat metric. But maybe that's just a technical complication, and we can somehow do the moral equivalent of keeping a vector constant as we move it.

We can. **Parallel transport** is the process by which we can start with a vector defined at some point and move it down a specified path, doing our best at each step to keep it parallel to its previous orientation. (As you might guess, the steps will be infinitesimally small, so we'll use calculus to chug the vector from point to point.)

But a subtlety arises. In flat space, not only did it seem intuitive, what we meant by "keep the vector constant," but it also didn't matter what path we took to move the vector from place to place. In a general curved space that's not going to be true, as we can see by thinking about parallel transport on a two-dimensional sphere.

Start at a point on the equator, and with a vector pointing north. Walk up to the north pole, doing your best to keep the vector constant; it's not that hard, as the vector will keep pointing in our direction of motion. But then imagine instead that we had walked for a while along the equator, also keeping the vector constant. In that case we would keep the vector pointing north. Then after some distance, we make a right-angle turn and head up to the pole, taking our vector with us.

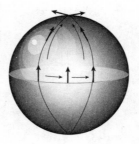

These two different routes from our point on the equator to the north pole have left us with two vectors pointing in obviously different directions, even though we did our best to keep the vectors constant the whole time. That wouldn't have happened in flat space, but on a sphere it's inevitable: Parallel transport along different trajectories generally leads to different outcomes. As we'll see in a moment, this failure of the two paths to give us the same results provides a nice way to formalize what we mean by "curvature." (Note that this operation is entirely intrinsic to the sphere—we don't have to look at it from the outside in order to parallel-transport a vector.)

We're bumping into a deep, and sometimes hard to internalize, feature of curved space (or spacetime): There is no unique way to compare vectors at different points. We can parallel-transport a vector along some particular curve, but using a different curve might give us a different answer. This means, for example, that we can't really talk about "the velocity" of a distant galaxy in an expanding universe. We do talk about that all the time, but implicitly we're making certain arbitrary choices about how to compare the vectors. That's okay, but we should keep in mind what is unique and well defined versus what just seems convenient. This is a version of the same lesson we preached in the context of the twin thought experiment in Chapter 6—it's best to think locally and compare quantities defined at the same point rather than to fool ourselves into comparing things going on far away.

GEODESICS

At the beginning of Chapter 3 we mentioned two ways of constructing a straight line between two trees. One was to stretch a string and pull it taut—a down-to-earth way of making a shortest-distance path—and the other was to just keep walking in a fixed direction. Both methods yielded the same result. And that continues to be the case on a general curved manifold in Riemannian geometry, although when space is curved it might not be natural to think of the resulting trajectory as "straight." On a sphere, for example, these paths are great circles or pieces thereof.

A path that minimizes the distance between two points (or maximizes the proper time, once we get to spacetime) is known as a **geodesic**. Geodesics obey an equation (Appendix B) that can be derived in a very similar way to how we thought about the principle of least action back in Chapter 3. There, we considered every path a particle could take, associated an action with each possibility, and demanded

that the action have zero derivative in the space of possible paths. Finding geodesics is precisely analogous, except we're minimizing the length of a curve rather than the action of a particle trajectory.

In addition to being shortest-distance paths, geodesics are also what you get if you just keep walking in a straight line. The formal version of that idea makes use of parallel transport. Consider a path, specified by the sequence of points making up the path, and also a parameter that tells us where we are along the path. We could write $x^i(t)$, for example, where x^i are the coordinates in however many dimensions we are in, and t is the parameter along the path. (Often the parameter really is "time," but here we're just using t as a convenient variable.) Then there is a vector, the velocity $v^i = dx^i/dt$, that points along the direction in which we are moving, with a length given by our speed.

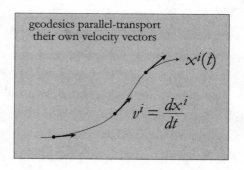

To "keep moving in the same direction" is to keep our velocity vector as constant as possible as we travel. That's just what parallel transport does. So another way of defining a geodesic is as "a path whose velocity at each point is just what we get by parallel-transporting the initial velocity vector." That's how, at the end of the day, the process of parallel transport is related to the metric tensor: Curves that parallel-transport their own velocity vector must also be curves of minimum length.

CURVATURE

Here's where we are: The metric tensor is the most basic geometric structure a manifold can have. It lets us calculate the distance along a path, areas and volumes of higher-dimensional regions of space, and the inner product between vectors. It also tells us how to parallel-transport vectors along a curve. That provides a nice way of thinking about geodesics—as curves that parallel-transport their own velocities in addition to being shortest-distance paths. But parallel transport is also crucial to the final piece of the puzzle: a complete characterization of the ways in which a space can be curved.

The sphere and the hyperbolic plane are both curved manifolds, but their curvature is as simple as it can be; the curvature looks the same at every point and in every direction. We'd like to be able to write down a robust, local characterization of how much curvature a manifold has at each point. The metric, as we've seen, isn't perfect for the job all by itself, since it can look simple or complicated in different coordinate systems even if the underlying geometry is the same. We'd like to define a quantity—a tensor of some sort, perhaps—that immediately reveals how curved a space is; one that is zero when space is flat and not zero when space is curved.

When we parallel-transported a vector along two different paths on a sphere, we ended up with two different vectors at the north pole. Equivalently, we could have started at the north pole and worked our way down to the equator, over, and back, traveling around a closed loop. This is the crucially important insight: If we parallel-transport a vector around a closed loop, it won't necessarily line up with the original vector upon its return. At least, not if the space through which it's moving is curved.

This suggests a way to characterize the existence of curvature. In flat space, parallel transport around a closed loop will always give us the same vector we started with. In a curved manifold, the vector might rotate along the way.

The problem is, there are too many closed loops. Nobody is going to easily write down the ways that vectors change when transported around every conceivable closed path. What we want is to pick out a special set of loops, ones we can easily characterize with just a handful of numbers.

We can address this problem with what is by now our usual trick: to think infinitesimally, and then use calculus to build up finite-sized results if desired. It's for this reason that the study of arbitrary curvatures on manifolds is called **differential geometry**.

Here's how we make a tiny loop to go around. Pick two vectors, $\vec{U}$ and $\vec{V}$, both defined at the same point p. Starting at p, travel an infinitesimal distance in the direction of $\vec{U}$. From there, travel an infinitesimal distance in the direction of $\vec{V}$. (Technically we move a distance that is proportional to the length of $\vec{U}$ and $\vec{V}$ in each case.) Then go backward from our original move, in the direction opposite to $\vec{U}$, and finally backward again in the direction opposite to $\vec{V}$. We thus return to our original point p, having described a tiny parallelogram along our journey.*

---

* If you're worried that curvature will lead us to slightly miss coming exactly back to our starting point, you are correct. But if our parallelogram is sufficiently small, the mismatch is negligible compared to the quantities we care about here.

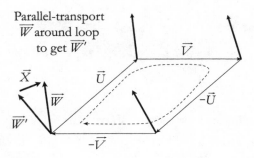

Parallel-transport $\vec{W}$ around loop to get $\vec{W}'$

That's a nice way to define a small loop using a relatively compact amount of information: just the components of two vectors at a point. To go from there to a measure of curvature, consider a third vector $\vec{W}$ at the original point. Now let's parallel-transport $\vec{W}$ around our tiny parallelogram, resulting in a new vector $\vec{W}'$. On a flat manifold, the vector won't change, and we will have $\vec{W}' = \vec{W}$. But on a curved manifold it might have shifted, if only a little bit. We can define the difference to be yet another vector:

$$\vec{X} = \vec{W} - \vec{W}'. \tag{7.20}$$

That's how we specify curvature at any point on an arbitrary manifold. Use two vectors to define a loop, parallel-transport a third vector around the loop, and a fourth vector representing the change in the third will be a measure of how curved the space is. If it's nearly flat, $\vec{X}$ will be small; in a highly curved space, $\vec{X}$ will be relatively large.

What we have, in other words, is a map from a collection of three vectors $\left(\vec{U}, \vec{V}, \vec{W}\right)$ to a fourth vector, $\vec{X}$. But we already have a name for maps between collections of vectors: They are tensors. So what we really have is the **Riemann curvature tensor**; it inputs two vectors to define a loop and a third vector to be transported, and outputs a fourth vector to characterize the change around the loop.

$$\vec{U}, \vec{V}, \vec{W} \rightarrow \quad R(\,,\,) \quad \rightarrow \quad \vec{X} = R(\vec{U}, \vec{V}, \vec{W})$$

It might still seem a bit cumbersome to ask what happens to every single vector that loops around every single parallelogram defined by two other vectors. That's where the magic of components comes to save us. Think of each of our vectors in terms of components, so that we represent $\vec{U}$ by a set of components U^i, and likewise for the other vectors. The number of components for each vector equals the dimensionality of the manifold we are in.

Then, just as the information about the line element can be encoded in the components g_{ij} of the metric tensor, the Riemann curvature tensor can also be encoded in a set of components, written $R^i{}_{jkl}$. (The order of the indices matters, as does whether they are up or down.) This gives us an explicit way of representing the map from $\left(\vec{U}, \vec{V}, \vec{W}\right)$ to $\vec{X}$, using components:

$$X^i = R^i{}_{jkl} W^j U^k V^l. \tag{7.21}$$

Note the Einstein summation convention in action. On the right-hand side we are summing over all the values of the repeated indices j, k, and l.

Admittedly, there are a lot of components in $R^i{}_{jkl}$. Four indices, each of which takes on d values in a d-dimensional manifold, leaving us d^4 components in total. That's 81 components in three dimensions, 256 components in four dimensions, and the number goes up rapidly from there.

Fortunately there are symmetries of the Riemann tensor that make life just a bit easier for professional general-relativists. Different components are related to one another in simple ways; for example, switching the last two indices gives the same number but with a minus sign: in $R^{i}_{\ jkl} = -R^{i}_{\ jlk}$. Thanks to this and other relations, the total number of independent components in d dimensions works out to be $\frac{1}{12}d^{2}(d^{2}-1)$. In four dimensions that's twenty independent components—considerable, but manageable. In two dimensions it's just one component. That means that in a two-dimensional manifold, there is simply a number at every point characterizing how curved the surface is (positively or negatively, as we know). In one dimension there are zero independent components! We think we can draw a one-dimensional curve that looks curved, but that's once again our extrinsic perspective fooling us. Inside a one-dimensional world, there would be no such thing as curvature.

The curvature depends on the metric in a well-defined way, but the dependence is complicated, so we have once again reserved the gritty details for Appendix B. What matters is that you understand the conceptual basis for what we've done: defining a metric, then parallel transport and geodesics, then curvature. Those will be the mathematical tools necessary to make sense of Einstein's general relativity.

EIGHT

GRAVITY

Classical mechanics, despite whatever impression you may have received over the previous chapters, is not a theory of physics. Rather, it's a *framework* within which honest theories can be constructed. Classical mechanics says, "A physical system is described by positions and momenta or appropriate generalizations thereof, and those variables obey equations given (equivalently) by Newton's $\vec{F} = m\vec{a}$ or Hamilton's equations or the principle of least action." It remains agnostic about what the forces are, what the specific Hamiltonian is, or what the action is. Newtonian gravity, by contrast, is a specific theory; it gives precise rules for what the forces are (the inverse-square law) from which definite predictions can be made. There are many specific theories under the broad umbrella framework of classical mechanics.

Quantum mechanics is an alternative framework to classical mechanics; there is a specific theory of the classical simple harmonic oscillator, and a separate theory of the quantum simple harmonic oscillator, for example. Almost all theories that modern physicists think about fall within either the "classical" or "quantum" frameworks.

Relativity is also a framework, not a theory. Specific theories can be "non-relativistic" (like Newtonian gravity) or "relativistic" (like Maxwell's theory of electromagnetism). The distinction comes down to how we think about spacetime: Non-relativistic theories feature absolute space and time and instantaneous action at a distance, while relativistic theories have light cones and a firm speed limit. The categories of non-relativistic/relativistic crosscut the categories of classical/quantum; there are models that fit in any of the corresponding categories.

	Classical	Quantum
Non-Relativistic	Newtonian Gravity	Quantum Harmonic Oscillator
Relativistic	Maxwell's Electromagnetism	Quantum Electrodynamics

It often happens in physics that we develop an understanding of some phenomenon in either a classical or non-relativistic context and then need to do a little work to construct an analogous understanding in the quantum and/or relativistic picture.

The ideas of relativity were largely inspired by features of Maxwell's electromagnetism, which remains the paradigmatic classical relativistic theory. But once relativity was in place, it was natural to turn to other forces of nature and ask how they could fit into the new framework. The obvious example to consider was gravity, which Newton had so successfully described right at the birth of non-relativistic classical mechanics.

Finding a successful relativistic theory of gravity turned out to be harder than anyone anticipated. It took a decade between Einstein's papers on special relativity in 1905 and his final formulation of general relativity in 1915. He wasn't the only one working on the problem; Finnish physicist Gunnar Nordström developed a competing theory, and he and Einstein corresponded as they worked. But Einstein's unmatched physical insight and gift for thought-experiment reasoning eventually won the day. His contributions to special relativity were enormously important, and it was his foundational work on quantum mechanics (also in 1905) that won him the Nobel Prize, but Einstein's proposal of general relativity as a theory of curved space-time was what made him the most celebrated scientist of the twentieth century.

INERTIAL AND GRAVITATIONAL MASS

Mass and gravity have a special relationship. Newton's second law says that the force acting on a body equals its mass times the acceleration it undergoes. Meanwhile, Newton's inverse-square law of gravity says that the gravitational force between two objects is proportional to the mass of each of them, and also to the inverse of the distance squared.

The idea of "mass" appears in both relations. But the role being played by that idea is completely separate in the two cases. In the second law, mass is a measure of inertia—the resistance that the body has to being accelerated. It's the same quantity no matter what force is causing the acceleration. But in the law of gravity, mass measures the interaction between the body and the specific force of gravity. Why are these two conceptually distinct quantities numerically equal to each other?

To emphasize this, let m be the inertial mass of a body, and M its gravitational mass. The second law is

$$\vec{F} = m\vec{a}, \tag{8.1}$$

while the gravitational inverse-square law for objects labeled 1 and 2 looks like

$$\vec{F} = G\frac{M_1 M_2}{r^2}\vec{e}_r.$$ (8.2)

Yet it just so happens that in the real world, the inertial and gravitational masses are equal to each other:

$$m = M.$$ (8.3)

There's no obvious reason it had to be that way. Consider the force between two electrically charged particles, given by Coulomb's law. It also takes an inverse-square form, and looks like

$$\vec{F} = K\frac{Q_1 Q_2}{r^2}\vec{e}_r.$$ (8.4)

This looks very similar to Newton's law of gravitation, with Coulomb's constant K replacing Newton's constant G, and electrical charges Q_1 and Q_2 replacing the gravitational masses. But there's a crucial difference. Unlike gravity, in electromagnetism the charge that acts as a source for the electric force has nothing at all to do with the inertial mass of the object. It can be positive (as in a proton), negative (as in an electron), or for that matter zero (as in a neutron). Different particles respond to the electromagnetic force differently, and some don't respond to it at all. Whereas every particle responds to gravity, exactly proportional to its mass, and gravity is always attractive.

Newton's theory of gravity doesn't attempt to explain why the mass that is the source of gravity is the same as the mass that governs inertia and motion. We just take it as a fact about the universe and move on.

THE PRINCIPLE OF EQUIVALENCE

The equivalence of inertial mass and gravitational mass has a provoca-
tive consequence. Consider two particles pulling on each other via
gravity, and let's think about what happens to particle 2 under the
influence of particle 1. Combining (8.1) and (8.2), we have

$$\vec{F} = m_2 \vec{a} = G \frac{M_1 M_2}{r^2} \vec{e}_r. \qquad (8.5)$$

But $m_2 = M_2$, so those quantities cancel out, leaving:

$$\vec{a} = G \frac{M_1}{r^2} \vec{e}_r. \qquad (8.6)$$

This is a formula for the acceleration felt by particle 2, which turns
out to be completely independent of particle 2's mass. In other words,
every object accelerates in the same way in a gravitational field, regardless of its mass (or anything else).

That's interesting. It certainly doesn't hold for other forces, like
electromagnetism. In a given electric field, a positively charged particle will be pushed in one direction and a negatively charged one in
the opposite direction. Gravity, by contrast, pushes everything in the
same direction, and at the same rate. This insight predates Newton's
equations, going all the way back to Galileo.

Einstein's breakthrough was to take this feature of gravity—all
particles fall at the same rate in a gravitational field—and extend it to
a general principle. In Chapter 3 we noted that if you were in a sealed
rocket ship, you could tell if you're accelerating or not accelerating,
but there is no way to gauge your absolute velocity. Now, let's instead
compare a rocket sitting motionless on the ground to one that is accelerating through space at one g (the acceleration due to gravity at the
Earth's surface).

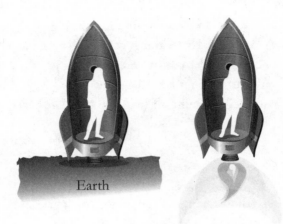

Earth

In the accelerating rocket, if you let go of an object, it appears to fall to the floor, due to the rocket's motion. All objects will fall in an identical fashion, since it's really the rocket that is to blame. But because of the equality of inertial and gravitational mass, the same thing happens in the rocket stationary on the ground. Any dropped objects will fall at the same rate, regardless of their mass.

This led Einstein to propose the **principle of equivalence**: In small regions of spacetime, the effects of gravitation are equivalent to those of being in an accelerating frame of reference. It's not just a matter of dropping objects to see how they fall; you can't do any experiment at all from inside your enclosed spaceship that would tell you whether you are accelerating through space or sitting peacefully on the ground. In operational terms: in small regions of spacetime, the laws of physics reduce to those of (non-gravitational) special relativity.

When we consider larger regions of spacetime, this clearly doesn't work. The gravitational field can pull in different directions depending on where you are located, whereas an accelerated frame will always lead to uniform motion for anything inside it. Near the Earth, for example, the gravitational force will always point toward the planet's center. If your spaceship were of a similar size to the Earth, you might notice that; widely separated pendulums would point in different directions. So the restriction of the principle of equivalence to small regions of spacetime is crucial.

You or I, having hit upon this insight, might be proud of ourselves and call it a day. Einstein, being Einstein, took it a step further. The fact that "physics looks like special relativity in small regions of spacetime" sounds suspiciously like something we've heard before. Just last chapter, in fact: Bernhard Riemann realized that the way to think about curved manifolds was to posit that "geometry looks Euclidean in small regions of space." So, Einstein reasoned, maybe we shouldn't think of gravity as a "force" living inside spacetime at all. Maybe we should think of it as a feature of spacetime itself: the curvature.

FREE FALL

Seems preposterous, doesn't it? Of course gravity is a force! Apples fall from trees, clumsy people fall down stairs, the Earth orbits the sun. How can we make sense of gravity other than to think of it as a force?

To be honest, you are still allowed to think of gravity as a force when it makes sense to do so—particle physicists like to talk about the four forces of nature, and they include gravity alongside electromagnetism and the strong and weak nuclear forces. But gravity is a different kind of force, precisely because it is universal. Gravity affects everything in the same way, unlike other forces that affect objects differently depending on their charges. That's what permits us to shift from thinking of a force propagating within spacetime to a property of spacetime itself.

Einstein's view demands a change of perspective. Say you drop a coffee cup and it falls to the floor. It's tempting to say, "the force of gravity pulled it down." But according to general relativity, the unforced motion of things is to be in free fall. The coffee cup is just doing its natural thing. It's you, standing on the ground, who are experiencing a force: The Earth is pushing upward on your shoes, deflecting you from a free-fall trajectory. It's the same set of events, just portrayed in a different way.

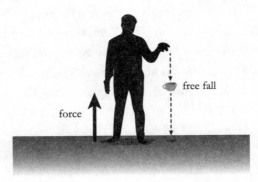

In special relativity, objects that don't feel forces move in straight lines. We know what the generalization of a straight line is to a curved manifold: It's a geodesic. According to general relativity, spacetime is curved, and unaccelerated objects move along geodesics of that curved manifold.

You might object that the Earth moves around the sun in something close to an ellipse, which is nobody's idea of a straight line. That's because you're still stuck thinking in space, rather than spacetime. In the sun's frame of rest, the Earth mostly moves in a timelike direction, since its motion through space is so slow (about 0.0001 times the speed of light). But the gravitational influence of the sun causes the spacetime around it to be gently curved, so that a geodesic around it looks like an ellipse in space. The Earth is simply doing its best to move in a straight line, just like the coffee cup.

THE SPACETIME METRIC

In Chapter 6 we talked about Minkowski spacetime and how the formulas for proper time and the length of spacelike curves look somewhat like Pythagoras's theorem, but with a funny minus sign. In Chapter 7 we talked about Riemannian geometry and how the fundamental idea is that of the metric tensor, which gives us a line element from which we can calculate the lengths of curves. These are closely related. Relativity is built on the idea that spacetime has a particular kind of metric, one with a minus sign for the timelike direction. A metric with this kind of minus sign is called **Lorentzian**, and if spacetime is perfectly flat we have the **Minkowski metric**.

Recall that there is a cute trick for indexing dimensions of spacetime, where we use Greek indices for all four spacetime dimensions (x^μ), then subdivide that into zero for time ($x^0 = t$) and Latin indices for space ($x^i = x^1, x^2, x^3$). With that notation, the Minkowski metric takes the form

$$g_{\mu\nu} = \begin{pmatrix} g_{00} & g_{01} & g_{02} & g_{03} \\ g_{10} & g_{11} & g_{12} & g_{13} \\ g_{20} & g_{21} & g_{22} & g_{23} \\ g_{30} & g_{31} & g_{32} & g_{33} \end{pmatrix} = \begin{pmatrix} -1 & 0 & 0 & 0 \\ 0 & +1 & 0 & 0 \\ 0 & 0 & +1 & 0 \\ 0 & 0 & 0 & +1 \end{pmatrix}. \quad (8.7)$$

In line-element form, this is equivalent to

$$ds^2 = -dt^2 + dx^2 + dy^2 + dz^2. \quad (8.8)$$

The minus sign leads to some awkwardness, but you learn to deal with it. The way we've written it, the length of a spacelike curve can be read off directly from the metric, which we essentially did back in equation (6.10). But for timelike curves—you know, the ones real objects <u>move</u> on—the spacetime interval ds^2 will be negative, so $ds = \sqrt{ds^2}$ appears to be imaginary, which is weird. That's okay; for timelike curves we simply deal instead with the proper time, which is defined by

$$d\tau^2 = -ds^2. \tag{8.9}$$

Alternatively you could simply multiply the entire right-hand side of (8.8) by –1, and define the line element to be minus what we've chosen it to be. Then the line element gives you the proper time directly and you have to multiply it by –1 when calculating spacelike distances. That's a respectable thing to do, it's just a choice of convention, and numerous textbooks work that way. (As a rough guide, physicists who specialize in relativity mostly use the $-+++$ convention we've chosen here, while particle physicists often go with $+---$.) Our choice is convenient if you occasionally like to think of "space at one moment in time," which inherits a nice Euclidean $+++$ metric this way.

If you wanted to boil special relativity down to a single compact sentence, it would be "spacetime has a Minkowski metric." The equations (8.7) and (8.8) contain all the information you need to talk about distances, times, light cones, space travel by twins or other adventurers, and more.

Einstein's insight was that the actual spacetime of our universe looks like Minkowski spacetime in small regions, but that the fabric as a whole is sewn together in such a way as to lead to curvature, as Riemann imagined. The entire geometry will be described by some kind of metric, but it won't be as simple as (8.7). Figuring out what the spacetime metric is and what influence it has on stuff in the universe is how professional general-relativists spend most of their working hours.

Let's look at a simple metric that is just a bit different from Minkowski. Here is the metric for an expanding universe:

$$g_{\mu\nu} = \begin{pmatrix} -1 & & & \\ & a^2(t) & & \\ & & a^2(t) & \\ & & & a^2(t) \end{pmatrix}. \qquad (8.10)$$

If a metric component is zero, we just leave it blank, trusting you to fill it in mentally if necessary. In line element form this is:

$$ds^2 = -dt^2 + a^2(t)\left[dx^2 + dy^2 + dz^2 \right]. \qquad (8.11)$$

The function $a(t)$ is called the **scale factor**. If we think physically about what's going on, we see that time is ticking away completely normally. We know that because the g_{00} component of the metric, the coefficient of dt^2, is -1 just like in Minkowski space. But the spatial parts are multiplied by the scale factor. If two objects, like galaxies, are located at fixed spatial coordinates, the distance between them will grow as $a(t)$ increases. Just like that, we have a mathematical description of expanding space.

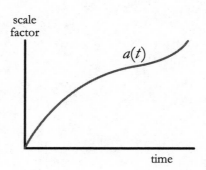

Thinking about general relativity often proceeds in the following way: We decide that we want to model a physical situation with given properties. For example, we might imagine that we have a universe uniformly filled with a smooth distribution of matter, rather than an inhomogeneous, lumpy configuration. Since the matter (which curves spacetime) is uniform through space, we can infer that the metric won't depend on spatial locations x^i, only on t. Furthermore, let's say that space (not spacetime) has a flat Euclidean geometry, rather than being positively or negatively curved. Those requirements lead us essentially uniquely to a metric of the form (8.10). But they don't tell us how the scale factor $a(t)$ behaves. That will depend on how much matter and other stuff there is in the universe. Clearly we need an equation that relates the metric to matter and energy. That's exactly what Einstein came up with.

THE ENERGY-MOMENTUM TENSOR

In mathematics, we write down axioms and rigorously derive theorems from them. In science, on the other hand, we basically guess—or "hypothesize," if you want to sound erudite—laws of physics, and then test them for both internal consistency and fit to experimental data. Einstein's task was to guess the equation that governs the behavior of the spacetime metric.

In Newtonian gravity, the force is proportional to the mass of the object causing the gravitational pull. We want a relativistic generalization of that idea. In relativity, "mass" is just one version of energy (it's the energy inherent in an object when it is at rest). And energy is unified with momentum—it's the zeroth component of the four-momentum vector, as we discussed in Chapter 6.

But that discussion was focused on the energy and momentum of a single object, which could be idealized as a point. We won't always be so fortunate. Sometimes we're going to want to consider sources of matter and energy that are spread throughout space, like the interior of a star or the distribution of dark matter around a galaxy. In relativity, such an extended distribution of matter is called a **fluid**. It's not a great term, since we often consider examples that aren't fluidlike at all. The interior of a planet is called a fluid, even if it's perfectly solid. A collection of photons moving in random directions is a fluid. To a general-relativist, "fluid" is any form of matter that is extended through space, rather than being concentrated at a point.

For a fluid, rather than thinking about the mass or energy, we can think of the **energy density**, usually written ρ (the Greek letter rho), the amount of energy per cubic centimeter (or whatever units you like to use). The energy density can vary across space and time, so in general it will be a function of x^μ. For a localized object like a star or planet, the total energy will just be the integral of the energy density over space.

But fluids are characterized by more than just the energy density at each point. There can also be **pressure**, which can be thought of as the amount of force the fluid would exert on the walls of a container. (Even if it's not literally confined in a container, we can think about the force it would exert if it were.) A fluid can be moving in complicated ways—think of currents in air or water—so in general there will be a velocity vector associated with the fluid at each point. And there can be other kinds of stresses and strains if the object is being twisted or deformed from its equilibrium shape. Like the energy density, all of these quantities will in general depend on what spacetime point we're looking at.

In relativity, all of these ideas are bundled up into the **energy-momentum tensor** (also known as the **stress-energy tensor**), usually written $T_{\mu\nu}$. The energy-momentum tensor summarizes everything we want to know about the mass, energy, momentum, pressure, stress, and other energylike features of a collection of matter (or radiation, or anything else).

In general, as you might imagine, an expression for $T_{\mu\nu}$ in terms of more familiar quantities will look fearsomely complicated. But we can get some intuition by considering the simple case of a **perfect fluid**: a fluid that looks the same in every direction in its rest frame. Then the only quantities we need to specify the energy-momentum tensor of that fluid are the energy density ρ and the pressure p. In flat spacetime, in the rest frame of the perfect fluid, the energy-momentum tensor looks like

$$T_{\mu\nu} = \begin{pmatrix} \rho & & & \\ & p & & \\ & & p & \\ & & & p \end{pmatrix}. \tag{8.12}$$

The 00 component is the energy density, and the diagonal spatial components are equal to the pressure. For non-perfect fluids or in non-rest frames, things get crazy; the pressure could be different in different directions, and there could be off-diagonal terms resulting from stress and strain. But we're already taxing our brains enough here, so it's okay to stick to the simple form (8.12). Both ρ and p can be functions of x^μ, so this expression can carry a lot of information. Perfect fluids are able to describe planets, stars, or dark matter and dark energy filling space.

EINSTEIN'S EQUATION

To generalize Newtonian gravity to the relativistic context, therefore, we want to invent an equation that relates the spacetime metric to the energy-momentum tensor. This is an extension of the unification we saw when we related the energy of a particle to its momentum. In general relativity, gravity isn't created just by mass, it's created by all different kinds of energy, pressure, stress, and so on.

So how do we do that? Both $g_{\mu\nu}$ and $T_{\mu\nu}$ are tensors with two lower indices, and as a bonus they are both symmetric ($g_{\mu\nu} = g_{\nu\mu}$ and $T_{\mu\nu} = T_{\nu\mu}$). So as a first guess, let's imagine they are proportional to each other:

$$g_{\mu\nu} = \alpha T_{\mu\nu}, \qquad\qquad (8.13)$$

where α is a constant of proportionality. Whenever you have an equation involving tensors, both sides need to have the same free indices in every term, otherwise you're not equating the same form of tensor.

This is a pretty silly guess, but we're trying to let you in on the kind of thing that would flicker across the mind of a theoretical physicist, even if for just a millisecond before being discarded. We immediately know it can't be right, because in empty space we should have $T_{\mu\nu} = 0$ (shorthand for 0s in every component slot), but we certainly don't want to have $g_{\mu\nu} = 0$. We want our equation to give us the Minkowski metric in empty space, or more specifically in situations where there is no gravity.

Let's use our brains a bit. Although (8.13) looks mathematically legitimate—it equates two symmetric two-index tensors—it doesn't make physical sense. Intuitively that equation would imply that energy-momentum is somehow *creating* the metric. That's not what we want at all. We want energy-momentum to *bend* the metric—not to create spacetime, but to curve spacetime. If there are no sources ($T_{\mu\nu} = 0$), spacetime can be flat, but if we stick a planet or star in there, spacetime must bend.*

If we think about it, what we really want is for the energy-momentum tensor to act as a source for *derivatives* of the metric, not the metric itself. Nonzero derivatives are the way we characterize bending. Back in Chapter 4 we mentioned how Laplace introduced the gravitational potential field as a way of thinking about Newtonian gravity. In that context the gravitational force depends on the

---

* Even with no sources, spacetime might not be flat—there might be gravitational waves passing through empty space, for example. But if spacetime is flat, there can't be any sources.

derivative of the potential, not on the potential itself. In our new relativistic context, we should think of the metric tensor as roughly analogous to the gravitational potential, whose derivatives give us the force, rather than the gravitational force itself.

So we're looking for a quantity that is a symmetric tensor with two lower indices (so we can set it proportional to $T_{\mu\nu}$) that we make out of the metric and derivatives thereof.

$$\left(\begin{array}{c} \text{symmetric tensor} \\ \text{constructed from} \\ \text{the metric and} \\ \text{its derivatives} \end{array} \right)_{\mu\nu} = \alpha T_{\mu\nu}. \qquad (8.14)$$

We almost have such a thing already in our possession. The Riemann curvature tensor $R^{\lambda}{}_{\sigma\mu\nu}$ (where now we're using Greek indices because we're in spacetime) is a tensor constructed from derivatives of the metric. The only problem is that it has too many indices. But there's another tensor, the **Ricci tensor**, that we can construct just by summing over the first and third indices of the Riemann tensor. The Ricci tensor is named after Italian mathematician Gregorio Ricci-Curbastro, who also invented the basics of tensor calculus and much of the apparatus of modern Riemannian geometry. On an influential review article, a 1900 paper with his former student Tullio Levi-Civita that Einstein would use to learn about tensors, for some unknown reason he signed his name "G. Ricci," without the "-Curbastro." All of his other papers included his full name; this was the only exception. Maybe he suspected that the tensor he would introduce would merit a short and memorable name.

Using Einstein summation convention, the Ricci tensor is

$$R_{\mu\nu} = R^{\lambda}{}_{\mu\lambda\nu} = R^{0}{}_{\mu0\nu} + R^{1}{}_{\mu1\nu} + R^{2}{}_{\mu2\nu} + R^{3}{}_{\mu3\nu}. \qquad (8.15)$$

We've changed around our Greek letters, but that's okay; they're just arbitrary index labels, after all. As long as our label choice is consistent (the same collection of free indices) throughout an equation, we should be fine. The Ricci tensor is also symmetric, $R_{\mu\nu} = R_{\nu\mu}$.

It sure seems like we are very close to the goal. We should obviously posit an equation of the form

$$R_{\mu\nu} = \alpha T_{\mu\nu}, \tag{8.16}$$

where α is again a constant of proportionality. This is a much more reasonable hypothesis than (8.13) was. It's of the general form (8.14), setting a symmetric two-index tensor constructed from the metric and its derivatives equal to the energy-momentum tensor. And in empty space, when $T_{\mu\nu} = 0$, it predicts that $R_{\mu\nu} = 0$, which is certainly consistent with flat Minkowski space (in which all components of the Riemann tensor vanish, so the Ricci tensor surely does).

It's so reasonable, in fact, that Einstein himself put forward this equation as a possible foundation for general relativity in October 1915. And it almost works, but not quite.

The problem is that we know something about energy: It's conserved. The question of conservation of energy in general relativity is a subtle one because energy can be transferred back and forth between matter and the curvature of spacetime. But once we take this into account, there is a strong restriction on how the energy-momentum tensor can change over time. And this restriction is not obeyed by the Ricci tensor. So if (8.16) were the correct equation, either energy isn't conserved at all, or almost no spacetime metrics would actually solve the equation.

Happily, there is an easy fix. Sadly, the details require a bit more digging into the nitty-gritty of tensors and curvature, which we have relegated to Appendix B. The essential trick is that we can define an **inverse metric**, $g^{\mu\nu}$, which is related to the metric but has upper rather than lower indices. (If you know about matrices, it is literally the matrix inverse of the metric.) Using the inverse metric we can define a function of spacetime, the Ricci curvature scalar, via

$$R = g^{\mu\nu} R_{\mu\nu}. \qquad (8.17)$$

This has no free indices at all, since we've summed over both μ and ν on the right-hand side. But we can multiply it by the metric $g_{\mu\nu}$ to obtain a separate symmetric two-index tensor constructed from the metric and its derivatives. Then, if you are Einstein scribbling feverishly in November 1915, you can try to figure out a combination of $R_{\mu\nu}$ and $Rg_{\mu\nu}$ that has the right properties to be proportional to $T_{\mu\nu}$ without violating conservation of energy. There is a uniquely right answer, which we nowadays call **Einstein's equation**:

$$R_{\mu\nu} - \frac{1}{2} R g_{\mu\nu} = 8\pi G T_{\mu\nu}. \qquad (8.18)$$

The combination on the left-hand side is the **Einstein tensor**. We could invent a new symbol for it, but it's simple enough to express it this way, as a combination of the Ricci tensor and curvature scalar. This is the final form of the field equation for general relativity, presented by Einstein in a lecture to the Prussian Academy of Sciences on November 25, 1915.*

General relativity was summarized by physicist John Wheeler as "spacetime tells matter how to move; matter tells spacetime how to

---

* There's one other thing you could do: Add a term proportional to the metric itself, $\Lambda g_{\mu\nu}$, where Λ is a constant. Einstein explored that possibility in 1917, calling Λ the **cosmological constant**. Evidence for a nonzero cosmological constant was finally gathered in 1998, when astronomers discovered the acceleration of the universe. But the measured value of Λ is so small that we can usually ignore it unless we're doing cosmology.

curve." The first half of that maxim is enacted by the idea that free particles move on geodesics, and non-free particles (ones subject to a force other than gravity) deviate from geodesic motion in much the same way that non-free particles accelerate away from straight lines in Newtonian mechanics. The second half is enacted by Einstein's equation, which we can solve to tell us what the spacetime metric will be in any situation of interest. This equation has ended up correctly predicting the evolution of the universe, the existence of black holes, the propagation of gravitational waves, and other phenomena that Einstein had no inkling of at the time. That's the power of a good scientific theory: It knows much more than the people who first write it down do.

The way we've written Einstein's equation doesn't include some undetermined constant of proportionality, but rather the specific factor $8\pi G$, where G is the same constant that appears in Newton's law of gravitation. You can't derive that value by pure thought, or even by seeking consistency with other cherished principles like energy conservation; you have to appeal to experimental data. What Einstein did was to consider the "weak-field limit," in which gravity is weak and spacetime is almost flat but not quite. In those circumstances we know that a good theory of gravity had better reproduce Newton's inverse-square law, and in order to make that work out, the constant in Einstein's equation (8.18) must be $8\pi G$. What's amazing is that this equation, with its numerical constant set by measuring apples falling from trees and the motions of planets in the solar system, makes predictions for what happens in the first minutes after the Big Bang, and passes with flying colors.

ACTION PRINCIPLE

Back in Chapters 3 and 4 we saw how there are ways of formulating classical physics that look different, but are mathematically equivalent: Newtonian mechanics, Lagrangian mechanics, and Hamiltonian mechanics. General relativity is a classical theory, so we should not be surprised that it can be derived in multiple equivalent ways. Let's look at the Lagrangian route via the principle of least action; this turns out to be an especially convenient way of thinking about relativistic theories, as it is naturally suited to treating space and time on an equal footing.

In our original look at the action principle, we started with a particle described by a position x and a velocity $v = dx/dt$. We defined a Lagrangian L as a function of x and v, in particular as the kinetic energy minus the potential energy. The action is the integral of the Lagrangian over time,

$$S = \int L\left(x, \frac{dx}{dt} \right) dt. \tag{8.19}$$

The actual path taken by a real particle is the one that minimizes this action compared to other paths between the same starting and ending points.

Now we have a somewhat different situation. Rather than a particle with a location in space, we are interested in the dynamics of the metric tensor. General relativity is an example of a **field theory**, since the metric tensor $g_{\mu\nu}(t, x^i)$ is a field that has a value at each point in spacetime, rather than a particle that has a location somewhere. In a field theory, we define the Lagrangian by constructing a function called the **Lagrange density** $\mathcal{L}$, then integrating that over all of space:

$$L(t) = \int \mathcal{L}\left(t, x^i \right) d^3 x. \tag{8.20}$$

The notation $d^3 x = dx^1 dx^2 dx^3$ indicates that we are integrating over all three dimensions of space. When you integrate a function of spacetime (the Lagrange density) over space, you end up with just a function of time (the Lagrangian itself). The action is the integral of L over time, which is just the integral of $\mathcal{L}$ over spacetime:

$$S = \int L\, dt = \int \mathcal{L}\, d^4 x. \qquad \text{(8.21)}$$

We've already figured out Einstein's equation, but let's imagine we hadn't, and we were trying to find it using the principle of least action. Our task is clear: We need to guess the appropriate Lagrange density $\mathcal{L}$. It should be constructed from the metric and its derivatives, just as the Lagrange density for a particle is constructed from position and its derivatives (in particular, the velocity). But the good news is that we are trying to guess a scalar function—a tensor with zero indices, if you like—rather than guessing a two-index tensor to put on the left-hand side of (8.14). There are fewer ways to have zero indices on a tensor than two indices, which makes our life considerably easier.

In fact there is essentially one possibility: the Ricci curvature scalar R. That is the obvious thing to guess for the Lagrange density for the metric, $\mathcal{L}_{\text{gravity}} = R$, because there's basically no other choice available. We should also include a Lagrange density for matter, but we don't have to be explicit about that; it will depend on what kind of matter we're interested in. And we need to put Newton's constant G in there somewhere to get the strength of gravity to work out correctly. The final answer turns out to be

$$S = \int \left(\frac{1}{16\pi G} R + \mathcal{L}_{\text{matter}} \right) \widehat{d^4 x}. \qquad \text{(8.22)}$$

That's it! This is the action that, when we look for spacetime metrics that minimize it, says that they will obey Einstein's equation (8.18). There is one detail we have suppressed for simplicity, namely, that the "volume element" in the integral is altered a bit in curved spacetime, so we've written it as $\widehat{d^4 x}$ rather than simply $d^4 x$ as a reminder.*

The beauty of the action formalism should be evident: It was way

---

* If you must know, the correct volume element is $\widehat{d^4 x} = \sqrt{-g}\, d^4 x$, where g is the matrix determinant of the metric tensor.

easier to guess the right scalar Lagrange density than it was to guess the right tensor for Einstein's equation, and cherished principles like energy conservation pop out automatically rather than lingering as things we have to worry about and verify. Of course, you do have to be pretty mathematically agile to figure out that you should be thinking about the action principle in the first place, and then to chug through the manipulations (which we have wisely avoided here) to actually derive Einstein's equation from the action.

Einstein himself was pretty mathematically agile, but not as much as his colleague David Hilbert, who was one of the great mathematicians of the early twentieth century. ("Hilbert space" appears as a crucial concept in quantum mechanics.) In the summer of 1915, when Einstein was closing in on the final form of general relativity but wasn't quite there yet, Hilbert invited him to give a series of lectures at the University of Göttingen. The two talked at great length about curved spacetime, and Einstein even stayed at Hilbert's house. They continued an extensive correspondence after Einstein returned to Berlin. And in the end, they both derived equation (8.18) almost simultaneously, Einstein via the guess-and-revise method, and Hilbert through the slick action formalism.

It's a matter of scholarly dispute whether Hilbert actually derived the complete field equation a few days before Einstein did, and how much useful input Einstein received from Hilbert's correspondence; there are missing letters, papers being updated during the editing process, and the usual messy minutiae of history. What's clear is that it was Einstein who came up with the basic insight of thinking of gravitation in terms of the curvature of spacetime, and who first presented the final equation in public, based on sensible physical criteria. Physicists therefore refer to (8.18) as "Einstein's equation," and (8.22) as "the Einstein-Hilbert action." That's a relatively accurate correspondence between labels and proper historical credit, which scientists don't always achieve.

EMPIRICAL CONSEQUENCES

Unlike most physical theories, the development of general relativity was not driven primarily by a need to explain some puzzling experimental anomaly but by the search for theoretical consistency. Einstein had in mind some features of how gravity worked, such as the inverse-square law and the principle of equivalence, and he was pretty familiar with the basic structure of relativity. It was a matter of reconciling these theoretical demands with each other, which he eventually accomplished by positing that gravity is a manifestation of the curvature of spacetime.

But once that was done, and the field equation derived, it was time to turn back to the empirical: to make experimental predictions and go out and test them.

One testing ground was the **precession of Mercury's orbit**, which was a little bit of a cheat, since the observational situation was already well-known. Kepler posited that planets move in perfect ellipses, and Newton put forward a theory in which that happens naturally if there is a single planet moving undisturbed around a perfectly spherical sun. In the real world, gentle tugs from the gravitational fields of the other planets cause orbits to precess a little bit, slowly shifting the direction along which the axis of the ellipse is oriented. Plugging in numbers, the orbit of Mercury is predicted by Newtonian gravity to precess by 0.148 degrees per century.

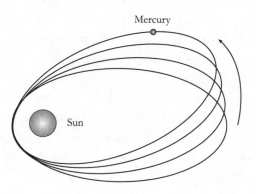

By the late 1800s, astronomers had measured the precession of Mercury and found 0.160 degrees per century. That's a discrepancy of 0.012 degrees/century—small, but large enough that it shouldn't be a random error. Urbain Le Verrier, a French astronomer who had proposed the existence of the planet Neptune to help explain the similarly anomalous orbit of Uranus, tried to repeat his success by suggesting a new planet orbiting inside Mercury. The hypothetical inner planet even got a name: Vulcan. Several teams claimed to have found it, but those claims all evaporated under closer scrutiny.

General relativity, Einstein realized, does a pretty good job at recovering the predictions of Newtonian gravity, but the match is not exact. There are small corrections that become increasingly important in stronger gravitational fields. We would therefore expect them to matter most, out of all the planets in the solar system, for Mercury, which is closest to the sun. Einstein set out to calculate the additional precession predicted by his theory, and found a result of 0.012 degrees/century, precisely the known discrepancy. One can only imagine Einstein's elation (and relief), after wrestling for years with tensor analysis and other mathematical abstractions, when he realized his theory provided a perfect explanation for a lingering observational anomaly.

It's great to solve a long-standing puzzle, but in science circles it's considered even more impressive if you can predict a phenomenon that hasn't yet been observed, and then go out and verify it. Such was the case for the **deflection of light**, or **gravitational lensing**, which Einstein predicted even before he had derived the full field equation for general relativity. That's possible because it's really a consequence of the principle of equivalence. From the perspective of an accelerating rocket ship, a light ray would appear to bend, just because of the changing motion of the ship. And if it's true for an accelerating rocket ship, it should be true for standing still on the surface of a planet with a gravitational field.

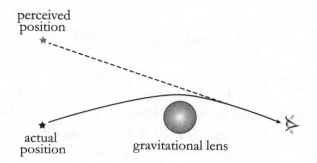

Or, even better, if light passes by the surface of a high-gravity object like the sun. The only problem is that the sun is pretty bright, and it's hard to see stars next to it, which we would have to do in order to measure their positions and determine whether the light from them has been deflected. This conundrum can be solved by waiting for a total solar eclipse, one of which conveniently took place in 1919. An expedition organized by British astrophysicist Arthur Eddington took photographs of stellar positions near the sun during the eclipse and verified Einstein's prediction of light bending.

It was this empirical result, more than his original theoretical proposal, that made Einstein into an international celebrity. The results of Eddington's observations were front-page news around the world, including *The New York Times*. As part of an elaborate multipart headline, the newspaper proclaimed about general relativity that "Sir Oliver Lodge says it will prevail, and mathematicians will have a terrible time." Not true. Mathematicians were delighted.

As were physicists and astronomers. Today, observations of gravitational lensing have matured into a high-precision science and a crucial instrument in the toolbox of the modern cosmologist. We look at the positions of many galaxies in the deep universe and use the statistical patterns of gravitational lensing to infer the existence of concentrations of matter, much of it "dark matter." General relativity is unambiguous that all forms of energy cause spacetime to be curved, which makes the deflection of light a uniquely useful method for mapping out the distribution of matter throughout space.

After these classic tests that were performed before or soon after the formulation of general relativity, many other phenomena came to be described by the theory: Light climbing away from a gravitating body will lose energy and have its wavelength shifted to larger values, known as the **gravitational redshift**. Matter in motion will cause ripples in the curvature of spacetime that propagate outward at the speed of light, known as **gravitational waves**. Dense collections of matter will collapse under their gravitational pull, creating regions of space from which light itself cannot escape—also known as **black holes**. And the universe itself will not sit still; filled with matter, space must either expand or contract, and in the 1920s Edwin Hubble established the **expansion of the universe**. All impressive implications of general relativity that have now been observed to exquisite precision.

It's a shame that Albert Einstein, who died in 1955, didn't live to witness the remarkable surge in the application of general relativity to important physical problems. During his lifetime, astrophysical observations generally weren't up to the task of investigating situations where relativity comes into play. Einstein himself never won the Nobel Prize for relativity, nor did Hubble for the expansion of the universe, in part because of a prejudice against astronomical questions.

Times have changed. Here is a complete list of Nobel Prizes that have been awarded for discoveries or phenomena where general relativity plays a central role, as of 2021:

- 1978 (Discovery of cosmic background radiation)

- 1993 (Binary pulsar, indirect evidence for gravitational waves)

- 2006 (Fluctuations and spectrum of cosmic microwave background)

- 2011 (Acceleration of the universe's expansion)

- 2017 (Direct observation of gravitational waves)

- 2019 (Evolution of galaxies and the universe)

- 2020 (Theory and observations of black holes)

As we can see, the pace is accelerating. Long considered an intellectual triumph but not really a central concern of the working physicist, general relativity is increasingly a central topic of exciting contemporary research. One likes to imagine Einstein would be pleased.

NINE

BLACK HOLES

Einstein's equation for general relativity packs a wealth of information into a compact package. We can thank the miracle of clever notation. The equation is meant to determine the metric on spacetime, $g_{\mu\nu}(x)$, but it's written in terms of the Ricci tensor, constructed out of the Riemann curvature tensor. Those tensors are defined in terms of the metric, to be sure, but if we were to write out that dependence in all its explicit glory, it would lead to a blizzard of terms that would fill an entire page's worth of mathematical symbols.

Einstein himself was sufficiently impressed, or intimidated, by the complexity of his own equation that he immediately resorted to approximation methods, such as the Newtonian limit, in order to derive experimental predictions. The equation seemed too complicated to be solved exactly, even in simplified situations.

That didn't deter Karl Schwarzschild. An accomplished astronomer and physicist, in 1915 Schwarzschild was serving in the German Army in World War I. He spent time on the French and Russian fronts, calculating missile trajectories. But while on temporary leave he was able to attend one of Einstein's lectures to the Prussian

Academy, and he became fascinated by general relativity. After returning to his war duties, in late December 1915 Schwarzschild was able to write Einstein a letter containing the first exact solution to Einstein's equation, describing the metric outside a spherical planet or star. Unfortunately Schwarzschild contracted a rare skin disease at the front, of which he died less than six months later at age forty-two. It took decades for physicists to come to terms with a mind-boggling and unanticipated implication of his discovery: the prediction of black holes in general relativity.

THE SCHWARZSCHILD SOLUTION

Schwarzschild was looking for the equivalent of Newton's inverse-square law in the solar system. In general relativity, that means finding a solution to Einstein's equation for the metric in the empty space around an isolated spherical body such as the sun. By calculating geodesics of that metric, we learn about planetary orbits, the deflection of light, and other predictions of general relativity. It would be entirely okay to simply present the Schwarzschild solution in all its glory, then to move on to discussing some of its implications. But it's a little more fun to work our way up to it, tracing the basic steps to illustrate how a theoretical physicist would go about tackling a problem like this.

We know that the metric on flat Minkowski spacetime, written in Cartesian coordinates (t, x, y, z), takes the form

$$g_{\mu\nu} = \begin{pmatrix} g_{tt} & & & \\ & g_{xx} & & \\ & & g_{yy} & \\ & & & g_{zz} \end{pmatrix} = \begin{pmatrix} -1 & & & \\ & +1 & & \\ & & +1 & \\ & & & +1 \end{pmatrix} \quad (9.1)$$

As before we haven't written out the 0s for the off-diagonal elements, but you know they are there.

A curved metric will have some (or perhaps all) of the metric components $g_{\mu\nu}$ be functions of some (or perhaps all) of the coordinates. That sounds like it could get very complicated very quickly. But we are enormously aided by the fact that the physical situation we care about (spacetime outside a spherical body) should itself be spherically symmetric. Operationally, that means we should expect the metric to depend on the distance from the origin, $r = \sqrt{x^2 + y^2 + z^2}$, rather than on $x, y,$ or z individually.

This suggests a first step: switching the spatial part of our coordinates from Cartesian (x, y, z) to spherical (r, θ, ϕ). We wrote down the metric for flat Euclidean space in spherical coordinates back in equation (7.12). From that it's straightforward to figure out what the metric on Minkowski spacetime looks like in (t, r, θ, ϕ) coordinates:

$$
g_{\mu\nu} = \begin{pmatrix} g_{tt} & & & \\ & g_{rr} & & \\ & & g_{\theta\theta} & \\ & & & g_{\phi\phi} \end{pmatrix} = \begin{pmatrix} -1 & & & \\ & +1 & & \\ & & r^2 & \\ & & & r^2(\sin\theta)^2 \end{pmatrix}.
$$

$$(9.2)$$

We just took the spatial part of the metric and replaced it with the spherical-coordinates version. To emphasize, this is still the metric on flat Minkowski spacetime, nothing to do with gravity yet. We're just writing it in coordinates chosen to help us work our way toward what the metric outside a gravitating body should look like.

In spherical coordinates, all we need to do is figure out how the metric depends on the radial coordinate r; the dependence on the angular coordinates (θ, ϕ) is fixed by spherical symmetry. Namely, the metric won't depend on ϕ at all, and it will only depend on θ through a factor of $(\sin\theta)^2$ in the $g_{\phi\phi}$ component in the lower right. And there are more simplifications. One is that we're looking for static solutions:

Spacetime is just sitting there, not evolving with time. So the metric components won't depend on t, either. We could also contemplate letting the off-diagonal terms be nonzero, but it's easier to guess something simple first, and in this case that works out for us.

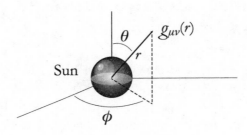

You might be disturbed by all this guessing. Seems unscientific. But when it comes to solving equations, guessing features of the solution is completely legitimate. We're not trying to find every single solution to Einstein's equation, just one particular case. At the end of the day, if our guess leads us to a metric for which we can calculate the Riemann tensor and Ricci tensor and show that it all satisfies Einstein's equation, none of the trickery we used to come up with that metric will matter.

Finally, let's look carefully at (9.2) and think about what the factors of r^2 are really telling us. They are there to reflect the fact that physical distances in the angular directions scale directly with the value of r. Here is a subtlety of general relativity: We're not really choosing coordinates and then solving for the metric. We're doing both things simultaneously. The coordinates don't have any meaning until the metric gives meaning to them, and the metric components make sense only with respect to some specified coordinates.

What this means is that we can simply *define* "r^2" to be "the quantity that appears in the angular components $g_{\theta\theta}$ and $g_{\phi\phi}$ of the metric tensor." Equivalently, we define r such that the area of a sphere at fixed

r is $A = 4\pi r^2$ and its circumference is $C = 2\pi r$. So we don't need to let those components be arbitrary functions and then solve for those functions. We're allowed to simply declare $g_{\theta\theta} = r^2$ and $g_{\phi\phi} = r^2 (\sin\theta)^2$ as already appears in (9.2), taking that as fixing the meaning of the coordinate r, as long as we are in a static, spherically symmetric situation.*

That leaves us with a metric of the following form:

$$g_{\mu\nu} = \begin{pmatrix} g_{tt} & & & \\ & g_{rr} & & \\ & & g_{\theta\theta} & \\ & & & g_{\phi\phi} \end{pmatrix} = \begin{pmatrix} -A(r) & & & \\ & +B(r) & & \\ & & r^2 & \\ & & & r^2 (\sin\theta)^2 \end{pmatrix}.$$

(9.3)

Not bad. All of our guessing based on shrewd physical intuition has left us with a fairly simple metric that has only two undetermined functions of one variable, $A(r)$ and $B(r)$.

Unfortunately, that's as far as shrewdness is going to take us. At this point we, or really Schwarzschild, have to suck it up and calculate the Riemann tensor and then the Ricci tensor. We won't do that explicitly, although all the technology necessary to do so is given in Appendix B, so feel free to give it a shot. For now we'll just accept that it can be done, and from there we can calculate the terms on the left-hand side of Einstein's equation. The resulting expressions will

---

* This choice means that r will not necessarily (or actually, as it turns out) be the distance from the origin. That's a different physical quantity, and we don't know ahead of time how it will relate to distance as measured along spheres of fixed radius. We can pick one, and then have to see what happens to the other. Area and circumference are both defined on the sphere itself, without reference to diving inside.

be written in terms of $A(r)$ and $B(r)$ and their derivatives with respect to r.

Then we have to set those expressions equal to the right-hand side of Einstein's equation, namely, the energy-momentum tensor. But more good news comes our way: At the moment we're only concerned with what the metric is outside the gravitating body, where space is empty and $T_{\mu\nu} = 0$. Complicated things might happen inside the sun or a planet, but for our present purposes those are just spherical sources causing spacetime to warp around them.

Schwarzschild did all of that, and to his surprise the resulting equations weren't that hard to solve. (To be honest, they weren't exactly trivial, but Schwarzschild was pretty smart.) Here's the expression he found for $A(r)$ and $B(r)$:

$$A(r) = \frac{1}{B(r)} = 1 - \frac{2GM}{r}. \tag{9.4}$$

The full Schwarzschild metric, in other words, is

$$g_{\mu\nu} = \begin{pmatrix} g_{tt} & & & \\ & g_{rr} & & \\ & & g_{\theta\theta} & \\ & & & g_{\phi\phi} \end{pmatrix} = \begin{pmatrix} -\left(1 - \frac{2GM}{r}\right) & & & \\ & +\left(1 - \frac{2GM}{r}\right)^{-1} & & \\ & & r^2 & \\ & & & r^2 (\sin\theta)^2 \end{pmatrix}. \tag{9.5}$$

Or in line-element form,

$$ds^2 = -\left(1 - \frac{2GM}{r}\right)dt^2 + \frac{1}{\left(1 - \frac{2GM}{r}\right)}dr^2 + r^2 d\theta^2 + r^2 (\sin\theta)^2 d\phi^2. \tag{9.6}$$

Isn't it gorgeous? That's the exact solution to Einstein's equation in empty space with spherical symmetry. In fact, it's the unique solution—we described how to guess it, but more rigorous analysis can prove that Schwarzschild is the only possible spherically symmetric metric that solves Einstein's equation in vacuum. You might object that Minkowski is another such solution, but just set $M = 0$ in (9.5) or (9.6) and you'll get Minkowski back as a special case.

In these expressions for the Schwarzschild metric, G is of course Newton's constant of gravitation, and M is the mass of the object whose gravitational field we have calculated. But just to be persnickety, we didn't derive the fact that M is the mass. What we derived (or sketched a derivation of) was the fact that for any M, a metric of the form (9.5) will solve Einstein's equation in vacuum. After the fact, we interpret M as the mass of some physical object like the sun, then we justify that interpretation by comparison with the Newtonian limit or something equivalent. It all works out in the end, but it's worth keeping in mind that concepts we take as self-evident generally depend on our background theoretical commitments, and they might undergo subtle shifts when we update our underlying theory.

TIME DILATION

Let's stare mindfully at the Schwarzschild metric (9.5) and try to figure out what it is telling us.

The point of a metric is to allow us to calculate distances in spacetime. That includes the proper time τ, which can be calculated by integrating $d\tau = \sqrt{d\tau^2} = \sqrt{-ds^2}$ along a timelike trajectory. Consider an object that stays fixed in the spatial coordinates (r, θ, ϕ). This is a good approximation to you and me standing on the surface of the Earth, or even the Earth in orbit around the sun; in both cases there is actually motion (the surface of the Earth is rotating, and the Earth is orbiting the sun), but the corresponding speed is very small compared to the speed of light and can be neglected at a first pass, in classic spherical-cow style.

For a path at constant spatial coordinates we have $dr = d\theta = d\phi = 0$. (That's what it means to be stationary—there is no increment in those directions.) Looking at (9.6), the proper time along such a path obeys

$$d\tau^2 = \left(1 - \frac{2GM}{r}\right)dt^2. \qquad (9.7)$$

This is pretty easy to integrate. Take the square root of both sides. The quantity in parentheses doesn't depend on t, so it's just a constant as far as time is concerned. For a finite interval we get

$$\Delta\tau = \int d\tau = \sqrt{1 - \frac{2GM}{r}} \int dt = \sqrt{1 - \frac{2GM}{r}} \Delta t. \qquad (9.8)$$

Remember that t is the coordinate time, a mere human convenience, while τ is the proper time, which is what clocks actually measure. This equation is telling us that a stationary observer will experience an amount of proper time that is proportional to the change in the coordinate time t, and the proportionality factor depends on the radial coordinate r.

That dependence is not hard to interpret. When r is large, the factor $\sqrt{1 - 2GM/r}$ is approximately 1, and the proper time and coordinate time are the same. Far away from the sun, the gravitational field is weak and spacetime is nearly flat. In those circumstances every fixed clock reads an amount basically equal to the coordinate time, much as Newton would have predicted.

But when r is close to (but still greater than) $2GM$, $2GM/r$ starts to approach 1, and $\sqrt{1 - 2GM/r}$ approaches zero. So any given change in t corresponds to less and less proper time as $r \to 2GM$. (For the moment, ignore what happens when r is less than or equal to $2GM$;

that's going to require more thought.) When the gravitational field becomes strong, proper time clicks off more slowly than coordinate time.

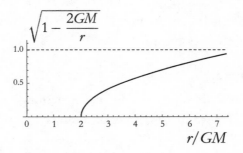

This is **gravitational time dilation**. We've been careful to emphasize that you shouldn't say "time runs more slowly in a gravitational field," but you can see why it's tempting. Clocks still run at one second per second, but their relationship to the coordinate time has changed. That doesn't matter to any fixed observer, since who cares about some made-up coordinate?

Where it does matter is if we compare what happens on two different paths between the same start and end points. Imagine you and your friend are far away from the sun and you synchronize your watches. Your friend stays behind while you travel closer to the sun, where you hang out for an extended period before eventually coming back. You rendezvous back with your friend at some value of the coordinate time, but your elapsed proper time will be less than that. You will have aged less, measuring less time on your clock. (In fact, the gravitational field is extremely weak even near the surface of the sun, by general-relativity standards, so this effect is pretty small.)

It's much like the twin in Minkowski space who zipped out near the speed of light, even if in this case you were moving slowly relative to your friend the entire time. It's the curvature of spacetime that is doing the work, not your velocity. Check out the movie *Interstellar* if you want to see this happen to Matthew McConaughey and Anne Hathaway.

Gravitational time dilation is a real physical effect, which has been verified experimentally. It's a vivid illustration of the idea that gravity really is geometry. In a very direct sense, gravity affects how clocks measure intervals in spacetime.

SINGULARITIES

From the above discussion, it seems that something special happens when the radial coordinate r reaches the value

$$r = 2GM. \tag{9.9}$$

This is indeed an important quantity, known as the **Schwarzschild radius**. Before getting too excited about it, let's admit that what happens at the Schwarzschild radius is completely irrelevant for the questions that originally motivated us—the spacetime outside a planet or star. That's because the Schwarzschild radius is extremely tiny. For an object the mass of the sun it's about three kilometers, while for an object the mass of the Earth it's less than a single centimeter. The actual sun has a radius of about 700,000 kilometers, and for the Earth it's a bit more than 6,000 kilometers. In either case the corresponding Schwarzschild radius is deep inside the gravitating body, where we're no longer in empty space.

The Schwarzschild solution, meanwhile, only applies to the metric on spacetime in vacuum, outside of the gravitating body. Once we dip inside, the metric is going to take on a different form, and nothing special at all is going to happen at $r = 2GM$. It would seem to be a mathematical curiosity, prompted by solving Einstein's equation in empty space and carelessly extrapolating the result to situations where space isn't empty at all.

But we're allowed to ponder what happens in a situation where space would be empty all the way down to the Schwarzschild radius and even smaller. What if, through some calamity, all of the mass of the sun were squeezed into a ball just a few kilometers across, or the mass of the Earth were squeezed into less than a centimeter? We'd have to imagine packing matter extremely densely indeed, but crazy things sometimes happen in astrophysics. As we'll see, it can and does happen, and the result is a **black hole**: a region of spacetime that is so dramatically curved that light itself cannot escape. For a black hole, where there is no matter to get in the way, the Schwarzschild radius defines the **event horizon**, the point past which it is impossible to return to the outside world.

As soon as Schwarzschild first presented his metric, people worried about what was going on at the Schwarzschild radius. Looking at the metric (9.5), we see that the factor $1 - 2GM/r$ goes to zero when $r = 2GM$, so the metric component g_{tt} vanishes. Interpreted physically, that would seem to imply that no proper time passes if we sit stationary at the Schwarzschild radius, even though the coordinate label t ticks away as usual. Hmm, that's weird, but maybe not too worrisome; it's somewhat reminiscent of what happens on any null trajectory, and light travels on such trajectories all the time.

More problematic is the g_{rr} component of the metric, which is $1/(1 - 2GM/r)$ and therefore blows up to infinity at $r = 2GM$. When a quantity becomes infinite, we say we are faced with a **singularity**.

That sounds bad. But is it really bad? After all, components of the metric depend on what coordinate system we have chosen. It's really the curvature that matters, not the metric components. Maybe we've just chosen an inconvenient coordinate system.

In fact, that's exactly what's going on, although it took physicists a number of years to figure it out. The apparent bad behavior of the metric at the Schwarzschild radius is merely a **coordinate singularity**—not anything physically ill-defined, just a bad choice of coordinates at that point. Any coordinate-invariant function you might construct out of the Riemann tensor remains finite at $r = 2GM$, even though a metric component blows up. There is something undeniably interesting happening at the Schwarzschild radius—it's the event horizon of a black hole, as we'll see in a moment—but spacetime is completely well-behaved there.

This line of investigation leads us to notice that the metric also seems to blow up at another location, namely, when $r = 0$ and therefore $2GM/r = \infty$. There, g_{tt} goes to infinity and g_{rr} goes to zero. We might hope that there's a similar story to tell as with the Schwarzschild radius, and maybe it's our coordinate system that is to blame.

No such luck. The location $r = 0$ is a true **curvature singularity**, where the curvature of spacetime itself seems to become infinitely big. That actually is bad. We might hope that this singularity came about only because we oversimplified the situation by assuming exact spherical symmetry, and perhaps it can be avoided in more realistically messy situations. This hope was dashed in a series of **singularity theorems** proved by Roger Penrose and Stephen Hawking in the 1960s, which demonstrated that curvature singularities are predicted to arise under a wide variety of physically realistic conditions.

General relativity plays games with us, however. According to the **cosmic censorship conjecture**, formulated by Penrose in 1969, any singularity predicted by general relativity will be hidden behind an event horizon. There are no **naked singularities** out there in the world, unclothed by horizons, for us to study up close. Modern numerical simulations seem to indicate that cosmic censorship is not exactly true, but is true almost all the time. You can make up initial conditions that evolve to a naked singularity, but those conditions have to be infinitely precise; any deviation results in the singularity being stuck behind an event horizon. Searching for naked singularities in the real world would not seem to be a promising research program.

Most physicists think that singularities don't really exist in nature, naked or otherwise. Their prediction is a sign that we've taken our theory too seriously in a regime where it should be expected to break down. General relativity is a classical theory, after all, and the world is fundamentally quantum-mechanical. We can hope that a true theory of quantum gravity will smooth out the singularities predicted by Einstein's classical theory, or at least resolve any conceptual puzzles they raise, but as yet we don't understand that hoped-for theory very well.

BLACK HOLES

The Schwarzschild radius defines a surface called the event horizon, and the region of spacetime inside the event horizon is a black hole. Let's dig into what that means, imagining a spacetime that is everywhere described by the Schwarzschild metric, with no extended object like a star or planet getting in the way.

When someone hands you a metric, a good way of getting a handle on what's going on is to look at the light cones. The set of all light cones, after all, is real structure on spacetime, not some person's favorite way of defining coordinates or slicing spacetime into space and time. So it will be illuminating to plot some light cones in a spacetime diagram of the Schwarzschild geometry.

Here is the answer, plotted in (t, r) coordinates; we'll contemplate it first, justify it afterward. Everything is spherically symmetric, which means that nothing especially interesting happens in the θ or ϕ angular directions. We've drawn both "ingoing" and "outgoing" light rays from a few points and only indicated future-oriented light cones. If the figure is confusing, don't let it get you down—it confused Einstein and other very smart people for decades. Afterward we'll change coordinates to better see what's going on, but it's useful to think in Schwarzschild coordinates first.

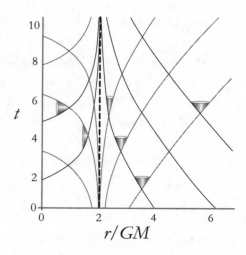

There's a lot going on in this figure. On the right, at large values of *r*, things look pretty much normal, with null rays tilted at 45 degrees and light cones pointing upward. That makes sense, since far away from the black hole there's no noticeable gravitational field and everything resembles Minkowski spacetime.

As we creep up on the event horizon at $r = 2GM$, the light cones begin to close up. That seems a bit strange. It's almost as if we can't cross the Schwarzschild radius, because the light cones won't let us. But perhaps it's just an artifact of our coordinate system going bad—as we'll soon see, that's exactly right.

It's inside the event horizon that things look crazy. The light cones start very wide, then grow narrower as we move toward smaller values of *r*. What's striking is that they point to the left, not up. But the rule that we have to stay inside our light cones when moving forward in time hasn't gone away. That means that *moving to smaller r, inside the event horizon of a Schwarzschild black hole, is moving forward in time.*

We've been wrong about $r = 0$ all along. Implicitly leaning on our intuition from flat spacetime, you've most likely been thinking of $r = 0$ as a location in space, the origin of coordinates at the center of a black hole. That's not right; $r = 0$ is not a location in space, it's a moment in time. Moreover, once you're inside the event horizon, it's a moment that is in your future. You cannot help but hit the singularity no matter how hard you might struggle. It's as inevitable as hitting tomorrow.

You might wonder how we can "discover" that $r = 0$ is a moment in time rather than a location in space. Aren't we the boss of our coordinates, and can't we define them as we choose? Certainly, but here we've already made our choice, back when we defined r in terms of the area of a sphere of fixed distance from the source. Once we made that choice—which we did in order to make things seem normal far away from the event horizon—we just extend the coordinate into a different regime and live with what happens. And what happens is that r becomes a timelike coordinate rather than a spacelike one.

It all seems quite bizarre. To make sure our feet are firmly planted in reality, let's dig into the equations responsible for the light cones we've drawn.

To "draw a light cone" means to pick a point and portray small line segments along which the spacetime interval is zero: $ds^2 = 0$. Glancing back at the line element (9.6), and ignoring θ and ϕ (because we're not moving in those directions), we have

$$-\left(1 - \frac{2GM}{r}\right)dt^2 + \frac{1}{\left(1 - \frac{2GM}{r}\right)}dr^2 = 0. \qquad (9.10)$$

Let's massage this a little bit. Move the second term over to the right-hand side, multiply both sides by –1 to get rid of the annoying overall minus sign, then divide by $(1 - 2GM/r)$. We get

$$dt^2 = \frac{1}{\left(1 - \frac{2GM}{r}\right)^2}dr^2. \qquad (9.11)$$

Looks like everything is squared, so let's take the square root of both sides, remembering to include a plus-or-minus symbol ± to indicate that either sign qualifies as a square root. Finally, divide both sides by dr to obtain

$$\frac{dt}{dr} = \pm \frac{1}{1 - \dfrac{2GM}{r}} \tag{9.12}$$

That's satisfying, because dt/dr is precisely the slope of the little line segment we draw to represent a light cone. (Remember we are specifically considering null trajectories, not arbitrary ones.) The plus/minus sign reflects the fact that light could be moving either inward or outward from the center.

Now we can connect this formula to the light cones we drew in the figure. At large values of r, we have $2GM/r \approx 0$, and therefore $dt/dr \approx \pm 1$. A slope of 1 or -1 indicates that our light rays are tilted at 45 degrees, which is just what happens in Minkowski spacetime, and indeed that's what we plotted in the figure.

As $r \to 2GM$, we get $1 - 2GM/r \to 0$, so $dt/dr \to \pm\infty$. That's the light cones "closing up"—their slope gets steeper and steeper on both sides, apparently coming together as we get closer to the event horizon.

Precisely at the Schwarzschild radius $r = 2GM$ the metric coefficients blow up, so let's put that aside for the moment. But if we fearlessly peek inside, something unexpected happens. When $r < 2GM$, the quantity $2GM/r$ will be greater than one, and therefore $1 - 2GM/r$ is going to be a negative number. From peering at the line element (9.6) we see that both g_{tt} (the coefficient of dt^2) and g_{rr} (the coefficient of dr^2) are going to change signs. Outside the event horizon, g_{tt} is negative and g_{rr} is positive, just like in Minkowski space. That reflects the fact that t is the timelike coordinate, while r is a spacelike coordinate.

But inside the horizon, these signs flip. This has a dramatic consequence: t is now spacelike, and r is timelike. Physically, once we're inside the Schwarzschild radius, moving to ever-smaller values of r is not moving toward the center, it's moving toward the future. Which is what physical particles actually do, of course. And why it's impossible to avoid ultimately reaching the singularity at $r = 0$.

This behavior is sometimes expressed (even by folks who really should know better) as "time and space swap roles inside the event horizon of a black hole." No. What happens is that the coordinates t and r swap their roles. Coordinates are a human invention and you shouldn't confuse them with fundamental features of reality. Time is still time and space is still space, no matter where we are or what coordinate system we are employing. If you were to fall into a sufficiently large black hole, you wouldn't even notice anything special upon crossing the event horizon. Your wristwatch would certainly not start measuring distance instead of time.

One more word of advice: if you fall into a black hole, don't struggle. You can't avoid the singularity in your future; in fact, you will hit it pretty quickly—a few hours for a billion-solar-mass black hole, and about one one-hundred-thousandth of a second for a one-solar-mass black hole. But that's if you just fall freely. And free fall, remember, is geodesic motion, which *maximizes* your proper time. If you accelerate in a foolhardy attempt to escape, you'll move on a path of shorter proper time. From your perspective, you'll just reach the singularity sooner.

THE EVENT HORIZON

The coordinate singularity at the Schwarzschild radius is the one ugly blotch on the picture we have just painted. From the spacetime diagram, as mentioned, one might wonder whether it's even possible to reach the event horizon, since the light cones appear to close up there.

A timelike trajectory would seem to just go straight up in the t coordinate, never actually crossing the Schwarzschild radius.

On the other hand, we learned from our look at gravitational time dilation that any fixed amount of t corresponds to less and less proper time τ as we approach the horizon. We can turn that around: Any fixed amount of proper time corresponds to more and more t as we approach the horizon. This is consistent with the idea that we're facing a coordinate problem rather than a physical one. Maybe, even though it looks like a timelike trajectory just shoots off to $t \rightarrow +\infty$, a real traveler could actually traverse an infinite amount of t in a finite proper time. So what happens to them a moment later?

What might solve this puzzle is a coordinate system that is better-behaved at the Schwarzschild radius, and happily there are such coordinate systems. A convenient one is called **Eddington–Finkelstein coordinates**, after Arthur Eddington and David Finkelstein. This system relies on the same spatial coordinates (r, θ, ϕ) that we've been using for Schwarzschild, but introduces a new time coordinate t^*, defined by

$$t^* = t + r + 2GM \log \left| \frac{r}{2GM} - 1 \right|. \qquad (9.13)$$

So our new time coordinate is our old time coordinate, plus our old radial coordinate, plus the logarithm of a certain function of the radial coordinate. It seems a little arbitrary, but there is a reason behind this specific form: Near the event horizon the logarithm goes to $-\infty$, which can cancel out t going to $+\infty$, and we can reach the horizon at a finite value of t^*.

Rather than showing that explicitly, it's easier to just write down the metric in the (t^*, r, θ, ϕ) coordinates and look at the light cones. The components of the metric are

$$g_{\mu\nu} = \begin{pmatrix} g_{t^*t^*} & g_{t^*r} \\ g_{rt^*} & g_{rr} \\ & & g_{\theta\theta} \\ & & & g_{\phi\phi} \end{pmatrix} = \begin{pmatrix} -\left(1-\dfrac{2GM}{r}\right) & 1 \\ 1 & 0 \\ & & r^2 \\ & & & r^2\left(\sin\theta\right)^2 \end{pmatrix},$$

(9.14)

which in line-element form corresponds to

$$ds^2 = -\left(1-\frac{2GM}{r}\right)\left(dt^*\right)^2 + dt^*\,dr + dr\,dt^* + r^2\,d\theta^2 + r^2\left(\sin\theta\right)^2 d\phi^2.$$

(9.15)

This is a bit different from what we've seen before; there is no g_{rr} component, but there are off-diagonal terms g_{t^*r} and g_{rt^*}. You might be surprised that g_{rr} disappeared even though we didn't change the definition of r; that's because the role of r has been partially absorbed into t^*, as can be seen in (9.13).

In the Eddington-Finkelstein coordinates, there is no more coordinate singularity at the Schwarzschild radius $r = 2GM$; all of the components in (9.14) remain finite numbers. (The $g_{t^*t^*}$ component goes to zero, but zero is a finite number.) This is reflected in how we draw the light cones, as portrayed in the next figure (with one angular dimension restored for aesthetic purposes).

In these coordinates, light cones don't close up at the horizon. What they do, instead, is tilt over as we move to decreasing r. Again we see that inside the horizon, staying inside the light cones forces us to move into the singularity at $r = 0$, which is in the future.

These improved coordinates help us understand what's happening at the Schwarzschild radius. The horizon is at a fixed radial coordinate, $r = 2GM$, and from the outside the black hole will look like a dark region of spacetime with a fixed size. But the light cones line up

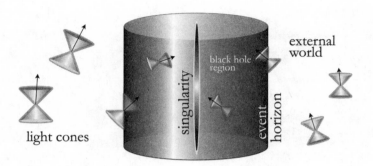

with the horizon on one side. At $r = 2GM$, the vertical t^* direction is actually null, rather than timelike (or spacelike).

That's what's so special about the event horizon. At $r = 2GM$, in order to stay on the horizon, you have to move in a specific direction, and do so at the speed of light. Any timelike trajectory would lead you deeper into the black hole. More generally, timelike paths can only pass one way through the horizon. It's not just that you can't imagine rocket engines powerful enough to escape from a black hole. It's that moving from the black-hole interior back to the outside world would require motion faster than the speed of light.

CHARGED AND SPINNING BLACK HOLES

The Earth is approximately spherical, but not quite: It is just a bit oblate, with the distance between the poles being shorter than the diameter along the equator by about 0.3 percent. Moreover, the Earth is lumpy, with deep oceans and tall mountains. As a result, the Earth's gravitational field is not quite uniform, and orbiting satellites have provided precise maps of exactly where the Earth's gravity is a bit stronger or weaker than average. Other planets will generally have their own features of local interest; the gravitational field of a planet can be highly distinctive.

Not so for black holes. They are not lumpy. On the contrary, there is an idea called the **no-hair theorem** (although mathematicians call it the "no-hair conjecture," since it hasn't been rigorously proven): Every black hole settles down to a state that is entirely characterized by its mass, electric charge, and spin. Any two black holes with the same values of those quantities will have the same gravitational fields. That's going to hold no matter what went into making them: A black hole created by the collapse of a massive star will end up looking exactly the same as one made from an equivalent amount of library books or peanut butter. The latter cases probably don't occur in nature, but if they did, you couldn't tell by examining the resulting black hole.

Black holes with electric charge but no spin will have a spherically symmetric electric field around them, much like a charged particle would. That electric field contains energy, which has to be accounted for when solving Einstein's equation. The metric for such a black hole is given by the **Reissner–Nordström solution**, which was derived soon after the Schwarzschild solution by a number of people. That's not so surprising; since everything is still spherically symmetric, the solution isn't all that different from Schwarzschild.

At least on the outside. The Reissner–Nordström solution, taken at face value and pushed as far as it can go, describes an infinite series of black holes connecting separate universes, all hidden behind the event horizon.

This is why you shouldn't always find exact solutions to equations and take them at face value, especially when extended beyond the physical situation they were intended to describe. Even the Schwarzschild solution, when maximally extended, describes two different universes connected by a **wormhole**. This fact was realized by Ludwig Flamm in 1916 and rediscovered by Einstein and his collaborator Nathan Rosen in 1935, so the Schwarzschild wormhole is sometimes called an **Einstein–Rosen bridge**.* But nobody believes that real

---

* I served as a science consultant on the 2011 film *Thor*. If you notice Jane Foster, played by Natalie Portman, talking about Einstein–Rosen bridges, that's my fault.

black holes out there in our galaxy are hiding wormholes and extra universes behind their event horizons. That's because real black holes aren't precisely empty space (or electric field) everywhere. They are formed by infalling matter, and the existence of that matter affects the spacetime metric in important ways. As a result, real black holes have singularities inside, but not gateways to other universes.

It took a lot longer to find the metric for a spinning black hole, which was eventually derived by Roy Kerr in 1963 and is now called the **Kerr solution**. The difficulty stems from the lack of spherical symmetry; there is a preferred direction, the axis of rotation of the black hole. But Kerr's work was much more than a mathematical tour de force. In the real world we don't expect black holes to have substantial electric charge; if one did, it would quickly be neutralized by attracting and absorbing particles of the opposite charge. But we do expect black holes to be rotating, and rapidly so. Almost every black hole in the universe will be accurately described by the Kerr metric.

THE AREA THEOREM

Although the idea of a black hole was implicit in Schwarzschild's work from 1915, it took a long time for physicists to understand the concept. The delay was due in large part to a difficulty in appreciating what features were physically real and what were merely coordinate artifacts, which is why we have been emphasizing that distinction. It was understood, for example, that gravitational time dilation became infinitely strong for an observer that hovered right at the Schwarzschild radius, so time would appear to freeze from the perspective of an outside observer. Therefore, people reasoned, an object that collapsed to that size would become a "frozen star," stuck forever at that radius. They didn't carefully ask what would happen from the point of view of an infalling observer, who we now know would simply dive right into a black hole.

It was in the late 1950s that event horizons were finally understood, due to work by David Finkelstein, Martin Kruskal, and others. This helped inaugurate a new era of active research in general relativity, led by John Wheeler in the United States, Dennis Sciama in the UK, and Yakov Zeldovich in the USSR. Wheeler popularized the term "black hole," and research on them was propelled forward by Penrose, Hawking, and their collaborators in the 1960s and '70s.

One important result of that period was the **area theorem**, proven by Hawking in 1971. It states that the area of the event horizon of a black hole, or the combined area of several black holes, only increases with time, never decreases. Like any good theorem, it relies on assumptions; for example, we assume that real particles have only positive masses, otherwise you could throw negative-mass particles into a black hole to make it shrink. And of course we are thinking purely classically—once quantum mechanics comes into the game, black holes will lose energy to radiation and therefore shrink after all, although it generally happens extremely slowly.

The area theorem becomes important when we're thinking about two black holes coalescing. When that happens, some of the total energy is going to be carried off in waves of gravitational radiation, since spacetime is being shaken by the rapidly moving black holes. But the area theorem guarantees that not too much energy will escape. The resulting single black hole will have an event horizon area that is at least as large as the sum of the areas of the two holes from which it was formed.

You might wonder why we don't simply say that the mass of a black hole never decreases, rather than saying that its horizon area never decreases, given that the Schwarzschild radius is $r = 2GM$ and the horizon area is therefore $A = 4\pi r^2 = 16\pi G^2 M^2$. The answer is that the mass *can* decrease, and that's because there are such things as spinning black holes.

For a rotating black hole, the horizon area depends on both the mass and the amount of spin. It turns out that energy can be extracted from a spinning black hole by throwing matter into it with opposite angular momentum. Roger Penrose devised an explicit way to do this, known as the **Penrose process**. We might imagine an advanced alien civilization that fed its power grid by extracting energy from a spinning supermassive black hole. But Hawking's theorem guarantees that the horizon area will increase all along the way, and eventually the black hole will no longer be spinning and there will be no further energy left to extract.

BLACK-HOLE MECHANICS

Something about the area theorem should bother you. It says that the area of a black hole event horizon will only increase over time. But that implies an arrow of time: The area increases toward the future, not to the past. Where did that come from? Einstein's equation doesn't treat the past and future any differently.

Part of the answer is a narrow and technical one. While general relativity doesn't by itself have an arrow of time, the black hole solutions we've been considering do. In the back of our minds, we imagine that there was a star or other astrophysical object in the past, which collapsed (as time moved toward the future) to make our black hole. It's possible to reproduce all of our mathematical discussion with the direction of time reversed. The result would be a **white hole**, containing a singularity in the past, surrounded by an event horizon from which matter can escape but never return. A white hole is a black hole run backward in time. We don't think that white holes exist in nature, but modern cosmology posits that our observable universe emerged from a Big Bang singularity in the past, and there is certainly a family resemblance there. The universe as a whole is something like a white hole.

This connection suggests a deeper relationship between black holes and thermodynamics (which is, after all, the origin of time's arrow). In the early '70s physicists pointed out that there was an intriguing correspondence between Hawking's theorem ("area always increases") and the second law of thermodynamics ("entropy always increases"). It was thought of as an amusing analogy, nothing more.

Until, that is, the work of Jacob Bekenstein, who was a graduate student working under John Wheeler. Bekenstein suggested that black holes really do have entropy, and that the entropy is proportional to the area of the horizon. More famous physicists scoffed at him, and for good reason: If black holes have entropy, then by the conventional rules of thermodynamics they would have to have a temperature. (In classical thermodynamics, when we change the energy of a system at fixed volume, the energy change divided by the entropy change is equal to the temperature.) And that means they would give off radiation. And that means they wouldn't be black after all.

Hawking, in particular, took umbrage at Bekenstein's proposal. But he took it seriously enough to set out to prove it wrong, using techniques that combined quantum field theory and general relativity. As is well-known by now, he ended up proving Bekenstein right. Black holes do have entropy, and they do have a temperature, and they do give off radiation, once quantum effects are taken into account. That's a story for another day, once we have some quantum mechanics under our belts. For now let us content ourselves with realizing that this **Hawking radiation** is extremely faint indeed. For a solar-mass black hole, the temperature is less than one-millionth of a degree Kelvin, much too small to be detectable (and larger black holes have lower temperatures).

THE REAL WORLD

It was at least a little bit of a surprise to physicists when the 2020 Nobel Prize in Physics was awarded to Roger Penrose, Reinhard Genzel, and Andrea Ghez. Penrose's share, in particular, represented a departure for the Royal Swedish Academy of Sciences. Nobody can doubt the brilliance or importance of his work, but Nobels are more often given for experimental findings or very specific theoretical models, not for proving that an established theory (general relativity) implies the existence of a particular phenomenon (black holes). But Penrose's case was enormously strengthened by the work of Genzel and Ghez, who had collected strong observational evidence for the existence of a real black hole at the center of our galaxy.

Black holes have moved from theoretical curiosities to the forefront of modern astrophysics. We haven't seen one up close—perhaps that's for the best—but there is overwhelming evidence that they exist and play important roles in multiple astrophysical processes.

Indeed, there is a whole menagerie of black holes with different sizes and origin stories. Perhaps the most famous kind are those formed at the end of the life of a massive star. Stellar energy comes from the fusion of light elements into heavier ones in the star's core. Eventually a star will exhaust all its usable fuel. At that stage, most stars will settle down to a **white dwarf** stage, which will gradually cool over time. More massive stars will contract to a **neutron star**, in which protons and electrons have combined into neutrons. The most massive stars, however, will collapse all the way down to black holes. We expect that most black holes formed in this way will be at least three times the mass of our sun.

One might wonder how we would ever detect the existence of stellar-remnant black holes. They are black, after all. But they are also spinning; even a small amount of rotation in the initial star can become significant when the mass is squeezed down to black-hole size. As a result, matter that gets attracted to the black hole tends to accumulate in an **accretion disk** in its equatorial plane. There can be a considerable amount of such matter, especially if the black hole is part of a binary system with another star. The temperature of matter in an accretion disk becomes very high, enough to give off high-intensity X-ray radiation. It's those X-rays that astronomers can observe. Most black holes won't be surrounded by conveniently observable accretion disks, but given the number of massive stars in the Milky Way, astronomers estimate that there could be hundreds of millions of stellar-mass black holes scattered throughout our galaxy. (There are about 100 billion stars here, so black holes are still a tiny fraction.)

There is an entirely different population, the supermassive black holes lurking at the centers of galaxies. These are millions or billions of times the mass of the sun, and astronomers believe they are present in most large galaxies in the universe. Our own Milky Way features a black hole of about 4 million solar masses. We have learned about it in part by tracking the high-velocity orbits of stars around a compact, dark region in the constellation Sagittarius. It is for those observations that Genzel and Ghez received their share of the Nobel Prize.

The Milky Way is a mature galaxy, in which most gas and dust has been converted to stars. This doesn't leave a lot of stray matter to fall into the central black hole, so ours is relatively quiet. But young galaxies often feature black holes with enormous, brightly shining accretion disks. We see these scattered throughout the observable universe as **quasars** or (more generally) **active galactic nuclei**.

Kip Thorne tells the story of the First Texas Symposium on Relativistic Astrophysics, held in Dallas in December 1963.* Astronomer Maarten Schmidt had just measured the distance to a quasar for the first time, demonstrating that it is extremely far away. Given how bright it appears to us on Earth, this object must be extremely luminous. Astronomers at the symposium were bubbling with excitement, tossing back and forth new scenarios for what quasars might be and how relativity might be important for understanding them. Meanwhile, a young mathematician from New Zealand gave an esoteric ten-minute talk on a new solution to Einstein's equation for a spinning spacetime. It was largely ignored by the audience, many of whom

---

* Kip Thorne, *Black Holes and Time Warps: Einstein's Outrageous Legacy* (New York: W. W. Norton, 1994), chapter 9.

slipped out of the hall for a break. The speaker was Roy Kerr, and almost nobody realized that he had described the metric for a spinning black hole, which would prove to be the crucial ingredient in understanding quasars.

Since 2015 we've had a brand-new way of gathering information about black holes: **gravitational waves**. If gravity is the curvature of spacetime, a gravitational wave is a ripple of curvature propagating at the speed of light. Just like ordinary electromagnetic waves can be produced by rapid motions of charged particles, gravitational waves are produced by rapid motions of massive objects.

The problem is that gravity is a weak force, so such waves are hard to detect. In 2015 such a detection was announced for the first time by the Laser Interferometer Gravitational-Wave Observatory (LIGO), working in collaboration with the Virgo Observatory in Europe. LIGO consists of two observatories, each with a pair of four-kilometer-long evacuated tubes oriented at right angles to each other. Lasers travel through the tubes and bounce off mirrors at the far end. A passing gravitational wave will induce a tiny distortion of the spacetime along the tubes, altering the time it takes the laser beams to travel to the mirrors and back. The signal is very tiny indeed: A typical wave will cause the mirrors to oscillate by a distance smaller than the width of a single proton. It is no surprise that it took many years of design and construction work, and many millions of dollars, to build such a sensitive instrument. Nor was it any surprise in 2017 when the Nobel Prize went to Rainer Weiss, Kip Thorne, and Barry Barish for their work in developing LIGO.

The 2015 event resulted from the merger of a 36-solar-mass black hole with a 29-solar-mass hole. These two behemoths, about 1 billion light-years away, had been orbiting closely to each other for an unknown period. The orbital motion produced gravitational waves, and the holes inched closer together as the system lost energy into this radiation. The eventual coalescence event lasted a matter of seconds until the single remaining black hole quickly settled down (in accordance with the no-hair theorem).

Since then, LIGO and Virgo have detected dozens more such events. Most of them have involved black holes between ten and one hundred times the mass of the sun. Partly that's a function of what kinds of black holes are out there in the universe, and partly it's due to the specific wavelengths the detectors are sensitive to.

Modern astrophysicists are extremely excited by the prospects for what we might learn from this new window on the cosmos. We'll learn more about black holes, of course, but also potentially about the life cycles of stars, the structure of galaxies, and perhaps about the size and shape of the universe. Most of all, like all good scientists, we are ready to be surprised by something completely unexpected.

APPENDICES

APPENDIX A: FUNCTIONS, DERIVATIVES, AND INTEGRALS

In this appendix we run through some very common functions and manipulations thereof, in case you might feel like taking a swing at solving some of the equations we've talked about.

A quick notational comment: we often use letters from the end of the alphabet, such as x, y, z, to indicate variables—quantities that are left undetermined in an equation, perhaps to be solved for later. Letters from the beginning of the alphabet, such as a, b, c, typically stand for constants—quantities that have some particular value, even if we might not tell you (or know) what it is. And letters from the late-early alphabet, like f and g, are often used for functions: maps from one variable to another. So a very standard thing to write would be $f(x) = ax + b$, where x is a variable, a and b are constants, and $f(x)$ is a function of x. The idea of x being a variable is that this relation is supposed to hold no matter what the value of x is, whereas a and b are thought of as fixed quantities, even if we don't tell you what they are. It's a subtle distinction.

Of course, this is all completely a matter of convention, and in principle we can use whatever letters we like. Before too long we'll run out of alphabet entirely and have to resort to Greek letters.

DEFINITE AND INDEFINITE INTEGRALS

In our introduction to integrals in Chapter 2, we glossed over one important detail: An integral represents the area under a curve. But that only makes sense if we specify a beginning and an end for the region whose area we are describing. This leads to a distinction between a **definite integral**, where we specify the endpoints, and an **indefinite integral**, where the endpoints are left unspecified.

Consider a case where the integral of some function $f(x)$ is some other function $F(x)$. In other words,

$$\int f(x)dx = F(x). \tag{A.1}$$

That's the indefinite integral. We're clearly being a little sloppy, since we didn't specify the endpoints. Sometimes to be careful we add "plus a constant," to indicate that the exact value will depend on the region chosen.* But often we just assume you are keeping your wits about you

---

* Two mathematicians are in a bar, arguing over the mathematical competence of the general public. One is extremely dismissive, while the other insists that laypeople can be surprisingly knowledgeable. When the first mathematician goes to the restroom, the second calls over their waitress. "Listen," he says, "when my friend comes back, I'll ask you a question, and you reply 'One-third x cubed,' okay? Don't worry about what it means, just say 'one-third x cubed.' Got that?" The waitress nods, and practices the words as she walks away: "One-third . . . x . . . cubed . . ."

When the first mathematician returns, the second one hails the waitress once more. "Help us out, would you? My friend here doesn't think that many people know anything about math. So let me just ask you, what's the integral of x squared?"

"One-third x cubed," she announces proudly. The first mathematician is impressed, and admits that his friend might be right after all.

As the waitress walks away, she looks back with a smile and adds, "Plus a constant."

and know what to do with the expression for the integral. Through-out these books, by "the integral of a function" we generally mean the indefinite integral.

The definite integral lets us delimit the region of integration ex-plicitly, by specifying the beginning at the bottom of the integral sign and the endpoint at the top:

$$\int_a^b f(x)\,dx = F(b) - F(a). \tag{A.2}$$

So the definite integral is the difference of the values of the indefinite integral at the final point and the initial point. Let's see it in action.

CONSTANT FUNCTIONS

Consider an extremely simple function, namely, a constant, $f(x) = c$. Not much detail to go into, but we're starting gently here. The slope of a constant is flat, so it comes as no surprise that its derivative is zero:

$$\frac{d}{dx} c = 0. \tag{A.3}$$

The indefinite integral is proportional to x itself:

$$\int c\,dx = cx. \tag{A.4}$$

That means that the definite integral is proportional to the distance between the endpoints:

$$\int_a^b c\,dx = c(b - a). \tag{A.5}$$

This is illustrated in the figure for $c = 2$, $a = 1$, $b = 3$. The area under the curve is $2 \times (3 - 1) = 4$, as we might hope.

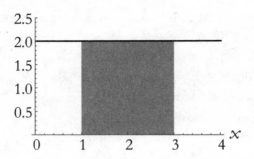

The parentheses in (A.5) mean that the quantity c is being multiplied by the quantity $(b-a)$, not that $b-a$ is the argument of a function, as x is in $f(x)$. Similar notation, different meanings; you're supposed to be able to discern the difference from context.

LINEAR COMBINATIONS

In mathematics we say that a sum of the form $af(x) + bg(x)$, where a and b are constants, is a **linear combination** of the two functions $f(x)$ and $g(x)$. The word "linear" refers to the fact that each function appears just once, and to the first power; multiplying functions together, or raising them to a different power, would be nonlinear.

Both differentiation and integration are **linear operators**. That means that the derivative of a linear combination is the linear combination of the corresponding derivatives, and likewise for integrals. So for derivatives we have

$$\frac{d}{dx}\Big[af(x)+bg(x)\Big]=a\frac{df}{dx}+b\frac{dg}{dx}, \qquad \text{(A.6)}$$

and for integrals,

$$\int\Big[af(x)+bg(x)\Big]dx=a\int f(x)dx+b\int g(x)dx. \qquad \text{(A.7)}$$

This works (of course) even if the second term is completely absent, and the function we want to differentiate or integrate is just $af(x)$. In these cases we can just "take the constant outside the derivative (or integral)." And if we're integrating with respect to x (as indicated by the dx symbol), then anything that doesn't depend on x counts as a constant, even if it depends on another variable: $\int f(x)g(y)dx = g(y)\int f(x)dx$.

PRODUCTS

Let's say we have the product of two functions, $f(x)\, g(x)$, where we'll henceforth drop the (x) and just write fg to keep things cleaner. Then there is a simple but somewhat nonintuitive formula for the derivative of the product:

$$\frac{d}{dx}fg = f\frac{dg}{dx} + g\frac{df}{dx}. \tag{A.8}$$

You can think of this as "first times derivative of the second, plus second times derivative of the first." It's called the **Leibniz rule**, after Gottfried Leibniz.

The existence of such an elegant procedure for taking the derivative of a product is the underlying reason why mathematicians generally consider differentiation to be "easy." Almost all functions we care about can be built up (perhaps recursively) in terms of combinations of other functions by addition, multiplication, and so on. The Leibniz rule implies that the derivatives of most functions can be written explicitly in terms of other functions (or "in closed form," as we say).

It would now be natural to present the formula for the integral of a product of two functions, but sadly no such formula exists. Integration is hard, both conceptually and in practice.

POWERS

Moving from general principles to specific functions, the most common thing we will encounter is a variable x raised to a power a, written x^a. The variable here is the **base** and the power is the **exponent**, but this is to be distinguished from the exponential function, where a constant is raised to the power of the variable, as we'll discuss below. If a is a positive integer, x^a is equal to x times itself a times. But there's no trouble using mathematical trickery to define x^a when a is not an integer, when it's negative, or even when it's a complex number.

Two useful facts about powers: when multiplying powers of the same variable, we simply add the exponents together, and when we raise a power to another power, the exponents are multiplied.

$$x^a x^b = x^{a+b}, \qquad \left(x^a\right)^b = x^{ab}. \tag{A.9}$$

Let's look at some simple (probably familiar) examples. The function x^2 is a parabola.

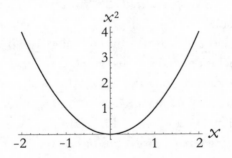

The value of this function is never negative, because multiplying two negative numbers together (x and itself) gives a positive number. The same goes for whenever we raise x to some other even integer, and the curve will look qualitatively the same. When the power a is a negative integer, the negative side of the function is minus the positive side, as in the case of x^3.

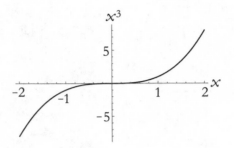

We can also define fractional powers, although in that case we're restricted to non-negative values of x. The way to think about it is that raising something to the power of $1/a$ undoes the action of raising it to the power a, because successive powers get multiplied in the exponent:

$$\left(x^a\right)^{\frac{1}{a}} = x^{\left(\frac{a}{a}\right)} = x^1 = x. \tag{A.9}$$

As a result, $x^{1/2} = \sqrt{x}$ looks like a parabola tilted on its side.

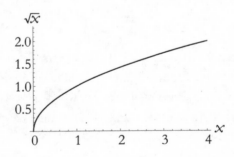

Negative or inverse powers can likewise be defined by thinking about the product of a power and its negative. The reciprocal function $x^{-1} = 1/x$, for example, can be defined by demanding that it satisfy

$$x \cdot x^{-1} = x^{1-1} = x^0 = 1. \tag{A.10}$$

The resulting graph has a discontinuity at $x = 0$, but it's nothing to be afraid of. We just say that $1/x$ is not defined at that point.

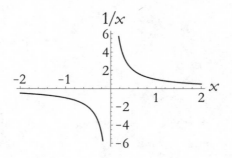

The derivative of a power is simplicity itself: lower the power by one, and multiply the entire expression by the original power:

$$\frac{d}{dx}x^a = ax^{a-1}. \tag{A.11}$$

The integral, sensibly enough, bumps up the power by one:

$$\int x^a dx = \frac{x^{a+1}}{a+1}. \tag{A.12}$$

For fun, you can verify that if you first take the derivative, then the integral, you get back to the original function, as you should.

There's a lurking problem, however: when $a = -1$, it looks like we're dividing by zero in (A.12). Indeed, and there's a special formula to handle that case:

$$\int x^{-1} dx = \ln|x|. \tag{A.13}$$

The vertical bars in $|x|$ are absolute value signs: if x is positive we leave it alone, if x is negative we multiply it by -1, so the result is non-negative. The function $\ln x$ is the natural logarithm, so we should explain that next.

EXPONENTIALS AND LOGARITHMS

We now move to the case where the variable is in the exponent, simply known as an **exponential** function, $f(x) = a^x$. The basic idea is pretty straightforward, with the plot shown in the figure for the case $a = 2$.

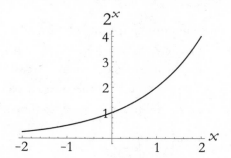

The inverse function of a^x is the **logarithm** (base a), which satisfies

$$\log_a\left(a^x\right) = a^{\log_a(x)} = x \qquad (A.14)$$

Just as exponentials are paradigmatic functions that grow quickly, logarithms (or just "logs") are paradigmatic functions that grow slowly as x gets large. The logarithm of 1 equals zero, and $\log_a(a) = 1$. At very small x the logarithm goes to $-\infty$, which makes sense if we think of $\log_a(x)$ as "the base we would have to raise a to in order to obtain x." Given a, to get a number close to zero by raising it to some power, that power has to be large and negative.

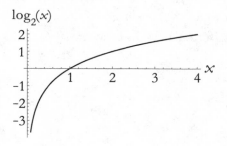

There is a wrinkle in the world of exponentials and logarithms: a special role played by a particular quantity, **Euler's number**:

$$e = 2.71828\ldots \qquad \text{(A.15)}$$

The expression for e contains an infinite number of terms that never fall into a repeating pattern. Like $\pi = 3.14139\ldots$ it is **irrational**, a number that is not expressible as the ratio of any two integers. There are many ways to define e, but perhaps the most satisfying is that e^x is the only non-constant function that is equal to its own derivative:

$$\frac{d}{dx}e^x = e^x. \qquad \text{(A.16)}$$

For exponentials with other bases, the derivative looks like

$$\frac{d}{dx}a^x = \ln(a)a^x. \qquad \text{(A.17)}$$

Here we've defined the natural log, which is just the logarithm base e:

$$\ln(x) = \log_e(x) \qquad \text{(A.18)}$$

It's the natural log that appears in (A.13), the integral of $1/x$. Looking at (A.17), and remembering that $\log_a(a)=1$ for any a, we see that the annoying factor of $\ln(a)$ disappears when $a = e$, leaving us with the elegant form (A.16). This is why most physicists use e as the base of their logarithms whenever possible.

The integral formula for an exponential is similarly straightforward,

$$\int a^x \, dx = \frac{a^x}{\ln(a)}. \qquad \text{(A.19)}$$

The derivative of a logarithm is

$$\frac{d}{dx}\log_a(x) = \frac{1}{\ln(a)}x^{-1}, \tag{A.20}$$

and its integral is

$$\int \log_a(x)\,dx = \frac{x\ln(x)-x}{\ln(a)}. \tag{A.21}$$

You can see how they all look their prettiest when $a = e$ and $\ln(a) = \ln e = 1$.

TRIGONOMETRIC FUNCTIONS

The last set of famous functions we'll look at are the trigonometric functions, especially the sine and cosine. In this case, the argument of the function is generally going to be an angle rather than a real number, so we'll denote it as θ rather than x. And importantly, we're going to be measuring our angles in radians rather than degrees. One hundred eighty degrees corresponds to π radians, so it's easy to convert back and forth.

We introduced trig functions back in Chapter 3, so here we skip right to some of their interesting properties. Pythagoras's theorem instantly gives us a famous relationship between the sine and cosine,

$$(\sin\theta)^2 + (\cos\theta)^2 = 1. \tag{A.22}$$

If we have a vector $\vec{v}$ with components v^i in flat, three-dimensional Euclidean space, we can define its **norm** (length) by Pythagoras once again,

$$|\vec{v}| = \sqrt{\left(v^1\right)^2 + \left(v^2\right)^2 + \left(v^3\right)^2}. \tag{A.23}$$

Then the dot or inner product between two vectors can be expressed in two equivalent ways, one using components and the other using the cosine of the angle between them:

$$\vec{v} \cdot \vec{w} = v^1 w^1 + v^2 w^2 + v^3 w^3 = |\vec{v}||\vec{w}| \cos\theta. \qquad \text{(A.24)}$$

One very nice thing about sines and cosines is that they are derivatives (and integrals) of each other.

$$\frac{d}{d\theta} \sin\theta = \cos\theta, \qquad \text{(A.25)}$$

$$\frac{d}{d\theta} \cos\theta = -\sin\theta. \qquad \text{(A.26)}$$

You just have to remember where the minus sign goes, which you can figure out if you remember that $\cos\theta$ is the one that starts at one and decreases, so its derivative for small positive angles has to be negative, which matches $-\sin\theta$. The integrals follow a similar pattern, with the minus sign moved (which makes sense if we remember that integrals undo derivatives).

$$\int \sin\theta \, d\theta = -\cos\theta, \qquad \text{(A.27)}$$

$$\int \cos\theta \, d\theta = \sin\theta. \qquad \text{(A.28)}$$

APPENDIX B: CONNECTIONS AND CURVATURE

In our discussion of geometry in Chapter 7, we covered all of the ideas necessary to understand the concepts of geodesics and Einstein's equation. But we didn't include absolutely all of the steps that would be necessary to calculate such things for a given metric. Here we fill in the gaps. We'll use Greek indices rather than Latin, imagining that we are in four-dimensional spacetime, but the formulas work equally well in space or in some other number of dimensions.

While working our way toward Einstein's equation in Chapter 8, we had to define the Ricci curvature scalar, which involved the idea of the "inverse metric." Let's be a little more explicit about what that means. We start by introducing an extremely useful tensor, the **Kronecker delta**, which has one upper index and one lower one. In four dimensions it looks like

$$\delta^{\mu}{}_{\nu} = \begin{pmatrix} 1 & 0 & 0 & 0 \\ 0 & 1 & 0 & 0 \\ 0 & 0 & 1 & 0 \\ 0 & 0 & 0 & 1 \end{pmatrix} = \begin{cases} 1, & \mu = \nu \\ 0, & \mu \neq \nu \end{cases}. \tag{B.1}$$

Thought of as a matrix, the Kronecker delta is just the **identity matrix**. The identity matrix plays the same role in matrix-land that the number 1 plays in ordinary arithmetic. When you multiply any matrix by the identity matrix, you get the original matrix back.

With that understanding, we can think of the inverse metric as the tensor we should multiply the metric by to obtain the Kronecker delta. The metric tensor $g_{\mu\nu}$ is a symmetric tensor with two lower indices, so the inverse metric will be a symmetric tensor with two upper indices, $g^{\rho\sigma}$, which satisfies

$$g^{\mu\lambda} g_{\lambda\nu} = \delta^{\mu}{}_{\nu}. \tag{B.2}$$

We should take a moment to appreciate what's going on with the indices here. There are two kinds of indices in a tensor expression: free indices and dummy indices. Dummy indices are the ones that appear twice, once upstairs and once downstairs, and are thus summed over, as with λ in (B.2). It doesn't matter what letter we use for them, as long as it's the same one, and there must be precisely one upper and one lower dummy index. (You can't sum over repeated indices that are both upper or both lower.) Free indices, on the other hand, appear only once in each expression, as with μ and ν in (B.2). The letters we choose to represent free indices also don't matter, but it is crucially important that they match up—each term (that is, each product of tensors) in any equation must have the same free indices. We see that in (B.2), where both the left-hand side and the right-hand side of the equation have an upper μ and a lower ν as free indices. If you are trying to add together tensor expressions with mismatched free indices, something has gone terribly wrong.

Ordinary Euclidean geometry implicitly uses a metric, they just don't tell you about it. The dot product between two three-dimensional Euclidean vectors, for example, is $\vec{v} \cdot \vec{w} = g_{ij} v^i w^j$. But the components of that flat Euclidean metric (in Cartesian coordinates) are

$$g_{ij} = \begin{pmatrix} 1 & 0 & 0 \\ 0 & 1 & 0 \\ 0 & 0 & 1 \end{pmatrix}. \tag{B.3}$$

Comparing to (the three-dimensional versions of) (B.1) and (B.2), the components of the inverse metric look precisely the same:

$$g^{ij} = \begin{pmatrix} 1 & 0 & 0 \\ 0 & 1 & 0 \\ 0 & 0 & 1 \end{pmatrix}. \tag{B.4}$$

This is why you may very well have experienced entire high school–level courses on geometry without ever hearing the word "metric." It was there, but it was always implied rather than spelled out explicitly, because you stuck with flat space in Cartesian coordinates and the components of the metric, inverse metric, and Kronecker delta were all the same.

That won't be true in general; components of the inverse metric are not typically the same as those of the metric. If the metric is diagonal, we're in the happy situation where the components of the inverse metric are simply the reciprocals of the metric components. (If the metric is not diagonal, things get messy quickly.) For example, in flat Euclidean three-dimensional space in spherical coordinates, the metric looks like

$$g_{ij} = \begin{pmatrix} 1 & 0 & 0 \\ 0 & r^2 & 0 \\ 0 & 0 & r^2(\sin\theta)^2 \end{pmatrix}. \tag{B.5}$$

This implies that the inverse metric has components

$$g^{ij} = \begin{pmatrix} 1 & 0 & 0 \\ 0 & r^{-2} & 0 \\ 0 & 0 & r^{-2}(\sin\theta)^{-2} \end{pmatrix}. \tag{B.6}$$

In flat space we at least have the option of using Cartesian coordinates where the metric and inverse metric are essentially the same, but in more general situations there is no such option, so it is important to keep the concepts distinct.

The presence of the metric and inverse metric enables a cute kind of tensor manipulation: **raising and lowering indices**. Whether a given index slot is a subscript or superscript matters, as the example of the metric itself shows. But we can turn an upper index into a lower one by multiplying by the metric and summing over that index, and likewise we can turn a lower index into an upper one with the inverse metric. Given a vector v^{μ}, for example, we can lower its index via

$$v_{\mu} = g_{\mu\nu}v^{\nu}. \tag{B.7}$$

The fact that the inverse metric satisfies (B.2) guarantees that we can lower an index then raise it again, returning to the same tensor that we started with (because summing over $\delta^{\mu}{}_{\nu}$ is the same as not doing anything at all):

$$g^{\mu\lambda}v_{\lambda} = g^{\mu\lambda}g_{\lambda\nu}v^{\nu} = \delta^{\mu}{}_{\nu}v^{\nu} = v^{\mu}. \tag{B.8}$$

It's this bit of tensor technology that was necessary to define the Ricci curvature scalar in Chapter 8. The Riemann tensor is naturally equipped with a single upper index and three lower ones, so it's straightforward to "contract" (sum over) one index to define the Ricci tensor, $R_{\mu\nu} = R^{\lambda}{}_{\mu\lambda\nu}$. But then we're stuck with a tensor with two lower indices; we can't do another contraction to get a scalar. What we can do is to raise an index using the inverse metric, $R^\mu_{\nu} = g^{\mu\lambda} R_{\lambda\nu}$. Then that can be contracted to define the curvature scalar, $R = R^\lambda_{\lambda}$, or equivalently $R = g^{\lambda\sigma} R_{\lambda\sigma}$. Then, if you're Einstein, you can put this to work defining a tensor that can be proportional to the energy-momentum tensor and still preserve energy conservation.

There's also a secret index-raising that was crucial along the way to defining the Riemann tensor in the first place. A central role was played by parallel-transporting a vector W^μ along a parameterized path $x^\mu(\lambda)$. (Despite being a Greek letter, λ here is not an index, it's the parameter telling us where we are along the path.) That means we have to define a value of the vector at each point, $W^\mu(\lambda)$, that satisfies the **equation of parallel transport**:

$$\frac{d}{d\lambda} W^\mu + \Gamma^\mu_{\sigma\rho} \frac{dx^\sigma}{d\lambda} W^\rho = 0. \qquad \textbf{(B.9)}$$

Most of the notation here makes sense, but we need to define $\Gamma^\mu_{\sigma\rho}$. These are known as **connection coefficients** or **Christoffel symbols**. They sure look like components of a tensor, but strictly speaking they are not, so we call them "coefficients" or just "symbols." (The reason is that they depend on coordinates in a non-tensorial way.) These coefficients define what we mean in practice by the **connection** on a manifold, the information we need to compare vectors and tensors at nearby points. The idea of a connection also plays an important role in gauge theories in particle physics.

To define the connection coefficients, we introduce yet more notation, but this time it's simply a labor-saving device rather than a novel conceptual twist. In the study of tensor fields on manifolds, it is extremely common to take partial derivatives with respect to the coordinates x^μ. So common that we invent a slick notation for just this purpose:

$$\frac{\partial}{\partial x^{\mu}} = \partial_{\mu}. \tag{B.10}$$

You see the trickery: x^{μ} has an upper index, but in the partial derivative it appears in the denominator, so the partial derivative operator ∂_{μ} comes with a lower index.

Now that we understand both the inverse metric and the partial-derivative notation, we can present the formula for the connection coefficients:

$$\Gamma^{\rho}_{\ \mu\nu} = \frac{1}{2} g^{\rho\lambda} \left(\partial_{\mu} g_{\nu\lambda} + \partial_{\nu} g_{\lambda\mu} - \partial_{\lambda} g_{\mu\nu} \right). \tag{B.11}$$

Chances are good that as you are reading this, somewhere in the world there are students learning general relativity who are calculating connection coefficients for some given metric using this equation. You're welcome to give it a go; using the flat metric in spherical coordinates (B.5) is hard enough to be interesting while still being tractable. Since there are three indices on $\Gamma^{\rho}_{\ \mu\nu}$, in three dimensions, that corresponds to $3^3 = 27$ components. But with a diagonal metric that only depends on two of the coordinates, many of those are going to end up being zero. Just keep in mind all the indices being summed over.

The connection coefficients define how parallel transport works, and therefore they also define geodesics, which after all are just paths that parallel-transport their own velocity vectors $dx^{\mu}/d\lambda$. Substituting this in for W^{μ} in (B.9), we obtain the **geodesic equation**:

$$\frac{d^2 x^{\mu}}{d\lambda^2} + \Gamma^{\mu}_{\ \rho\sigma} \frac{dx^{\rho}}{d\lambda} \frac{dx^{\sigma}}{d\lambda} = 0. \tag{B.12}$$

Given some metric $g_{\mu\nu}$, we can calculate the connection coefficients from (B.11), then use this equation to solve for geodesics $x^{\mu}(\lambda)$. That will tell us how real physical objects, from planets to photons, travel freely through such a spacetime.

Finally, the other important use of the connection coefficients is to define the Riemann curvature tensor. We explained it conceptually in Chapter 7, but at some point you're going to have to sit down and calculate the components. Here is the formula in all its glory:

$$R^{\rho}{}_{\sigma\mu\nu} = \partial_\mu \Gamma^{\rho}{}_{\nu\sigma} - \partial_\nu \Gamma^{\rho}{}_{\mu\sigma} + \Gamma^{\rho}{}_{\mu\lambda}\Gamma^{\lambda}{}_{\nu\sigma} - \Gamma^{\rho}{}_{\nu\lambda}\Gamma^{\lambda}{}_{\mu\sigma}. \quad \text{(B.13)}$$

The truth is that kids these days don't usually calculate the components of the Riemann tensor by hand. There are computer programs that will do it for you. They will never know the formative experience of spending late nights with pages of note paper scattered across a kitchen table, filled with Greek symbols, chasing down the step in which you mistakenly wrote μ instead of ν. Good times.

INDEX